Lecture Notes in Mathematics

Edited by A. Dold and B. Eckmann

Subseries: USSR
Adviser: L. D. Faddeev, Leningrad

1312

Nikolai A. Shirokov

Analytic Functions Smooth up to the Boundary

Springer-Verlag
Berlin Heidelberg New York London Paris Tokyo

Author

Nikolai A. Shirokov
Fontanka 27, LOMI
191011 Leningrad, D-11, USSR

Consulting Editor

Sergei V. Khrushchev
Fontanka 27, LOMI
191011 Leningrad, D-11, USSR

Mathematics Subject Classification (1980): 30D50, 46J20

ISBN 3-540-19255-7 Springer-Verlag Berlin Heidelberg New York
ISBN 0-387-19255-7 Springer-Verlag New York Berlin Heidelberg

Library of Congress Cataloging-in-Publication Data. Shirokov, Nikolai A., 1948- Analytic functions smooth up to the boundary. (Lecture notes in mathematics; 1312) "Subseries: USSR." Bibliography: p. Includes index. 1. Analytic functions. 2. Multipliers (Mathematical analysis) I. Title. II. Series: Lecture notes in mathematics (Springer-Verlag); 1312.
QA3.L28 no. 1312 [QA331] 510 s [515] 88-12336
ISBN 0-387-19255-7 (U.S.)

Printing and binding: Druckhaus Beltz, Hemsbach/Bergstr.
2146/3140-543210

CONTENTS

INTRODUCTION

1. This volume is mainly concerned with the Nevanlinna factorization in classes of functions analytic in the unit disc $\mathbb{D}$ and smooth in a sense up to the boundary $\partial\mathbb{D}$ (in what follows we call such functions smooth analytic functions due to the smoothness of their boundary values). Different kinds of factorizations (i.e. roughly speaking methods of decomposition of a function into the "simplest" factors) played an important role in complex analysis from the very beginning of its existence and even now continue to be a keystone of that branch of mathematics. The Weierstrass products in the theory of entire functions, the Blaschke products, the inner and outer functions form nowadays an essential part of the analytic machinery . The last three decades have provided new inventions in that field - we would like mention only a long series of papers by M.M.Dzhrbashian [40] , a new factorization of entire functions introduced by Rubel [42] , and the Horowitz products [43] . Interest in the different kinds of factorizations is stimulated by urgent problems of Complex analysis and first of all by the problems of uniqueness and of the distribution of values, which form the core of the subject.

Factorizations as a tool are widely used in the study of ideals in Banach algebras of analytic functions, in problems of spectral analysis and synthesis; their vector-and operator-valued analogues play a notable role in the modern spectral operator theory.

In the present volume we deal with the least known and most frequently used factorization, namely with the Nevanlinna factorization or, in modern terms, the inner outer factorization. Developed by R.Nevanlinna, G.Szegö and V.I.Smirnov, the factorization was intensively studied already in the 1920-s and 1930-s. Nevertheless the development of methematics during the last few decades has revealed an essentially new phenomenon which roughly speaking consists in the fact that the Nevanlinna factorization fits well not only to classes similar to the Hardy classes but also to classes of smooth analytic functions.

Let us recall some classical facts and notation (see [44] , [45] for details).

2. A function I analytic and bounded in $\mathbb{D}$ is called an inner function if $\lim\limits_{r\to 1-0}|I(r\zeta)|=1$ almost everywhere on $\partial\mathbb{D}$. Two important examples of inner functions are the following:

a) Let $\{\alpha_K\}$ be a sequence (perhaps finite) of points of $\mathbb{D}\setminus\{0\}$ satisfying

$$\sum_K (1-|\alpha_K|)<\infty .$$

Then the product

$$B=\prod_K \frac{\bar{\alpha}_K}{|\alpha_K|}\,\frac{\alpha_K-z}{1-\bar{\alpha}_K z}$$

converges in $\mathbb{D}$ to an inner function vanishing at the α_K's and only at them. Then the function $z^m B$, where $m\in\mathbb{Z}_+$, is called a Blaschke product.

b) Let μ be a nonnegative Borel measure on the circle $\partial\mathbb{B}$ which is singular with respect to Lebesgue measure on $\partial\mathbb{D}$. The function

$$S_\mu(\tau)=exp\Big(-\frac{1}{2\pi}\int_{-\pi}^{\pi}\frac{\zeta+\tau}{\zeta-\tau}\,d\mu(\zeta)\Big),\quad \tau\in\mathbb{D} \qquad (0)$$

is an inner function. It does not vanish in $\mathbb{D}$ and is called the singular inner function corresponding to the measure μ .

These two examples are the basic ones because any inner function may be uniquely factored into the product

$$I=cBS \qquad (1)$$

where $c \in \partial\mathbb{D}$, B is a Blaschke product and S is a singular function.

In what follows it is important to notice that as a rule I does not possess any smoothness on the circle $\partial\mathbb{D}$. In case I is continuous in $\bar{\mathbb{D}}$ we have $S \equiv 1$ and B is a finite Blaschke product.

c) Outer functions. Let $\log|h| \in L^1(\partial\mathbb{D})$. We can associate with h an analytic function $_e h$ in $\mathbb{D}$ (which is called the outer function corresponding the function $|h|$) as follows

$$_e h(\tau) = \exp\left(\frac{1}{2\pi}\int_{-\pi}^{\pi} \frac{e^{i\theta}+\tau}{e^{i\theta}-\tau} \log|h(e^{i\theta})|\,d\theta\right), \quad \tau \in \mathbb{D}.$$

The function $_e h$ does not vanish in $\mathbb{D}$ and

$$|_e h(\zeta)| = \lim_{r \to 1-0} |_e h(r\zeta)| = |h(\zeta)|$$

for almost all $\zeta \in \partial\mathbb{D}$.

d) Class N. A function f analytic in $\mathbb{D}$ is said to belong to the Nevanlinna class N if

$$\sup_{0<r<1} \int_0^{2\pi} \log^+ |f(re^{i\theta})|\,d\theta < \infty.$$

That class plays a very important role in analysis. Classes of analytic functions most frequently used in harmonic analysis and operator-theory are usually contained in N.

The following result is the starting point of the Nevanlinna's factorization theory.

THEOREM. Let B be a Blaschke product, S_μ be a singular function corresponding to a real Borel measure μ and $_e h$ be an outer function corresponding to the function h such that $\log|h| \in L^1$.

Then

$$f = cBS_{\mu e} h \in N \qquad (2)$$

Conversely, an arbitrary function $f \in N$ can be uniquely represented in the form (2). In what follows we write down the product in (2) as

$$f = {}_e f \cdot I_f$$

where $I_f = cBS_\mu$, ${}_e f = {}_e h$.

The importance of the Nevanlinna factorization is that it provides a complete description of the class as well as its crucial subclasses in terms of "pure real" parameters determining the factors I_f and ${}_e f$ (these are the constant $c(f) \in \partial\mathbb{D}$, the real measure $\mu = \mu(f)$), the sequence $\{\alpha_K\}$ and the number m and finally the values $|f|$ on $\partial\mathbb{D}$). The subclasses mentioned above are the Hardy classes H^p and the Smirnov class $\mathcal{D}$. We recall [44] that

$$H^p = \{f \in N : \sup_{0<r<1} \int_0^{2\pi} |f(re^{i\theta})|^p d\theta < \infty\}, \quad 0<p<\infty,$$

$$H^\infty = \{f \in N : \sup_{\mathbb{D}} \|f| < \infty\},$$

$$D = \{f \in N : \lim_{r \to 1-0} \int_0^{2\pi} \log^+ |f(re^{i\theta})| d\theta = \int_0^{2\pi} \lim_{r\to 1-0} \log^+ |f(re^{i\theta})| d\theta\}.$$

It can easily be cheked that $H^{p'} \subset H^{p''} \subset \mathcal{D}$, if $\infty \geqslant p' \geqslant p'' > 0$. An alternative description of $\mathcal{D}$ is the following: a function f belongs to $\mathcal{D}$ iff the singular measure $\mu(f)$ is nonnegative. Thus the Hardy classes can be characterized as follows:

$$H^p = \{f \in N : \mu(f) \geqslant 0,\ f|_{\partial\mathbb{D}} \in L^p(\partial\mathbb{D})\}.$$

3. Let us now dwell on some details connected with the formula (2) which are especially important for our paper.

I. We say that an inner function I_2 divides an inner function I_1 if $I_1/I_2 \in H^\infty$. The above references yield that if $f \in \mathcal{D}$ and if an inner function I divides I_f then $fI^{-1} \in \mathcal{D}$. Similarly if $f \in H^p$ and I divides I_f then $fI^{-1} \in H^p$. Roughly speaking the function fI_f^{-1} is obtained from f by "removing the zeros" of f. As a matter of fact the function fI^{-1} does not vanish in $\mathbb{D}$, since fB_f^{-1} has no zeros in $\mathbb{D}$ and removing of f_μ means (in a sense) an isolation of the "boundary zeros" of f. The outer factor f behaves in Approximation Theory and Theory of Invariant Subspaces in many respects as invertible. Thus $\mathcal{D}$ and H^p are invariant with respect to the "isolation of zeros".

II. Nevanlinna's theorem contains the complete information about moduli of functions from N or H^p on $\partial\mathbb{D}$.

For example, let h be a nonnegative function on $\partial\mathcal{D}$. Then the following statements are equivalent:

(α) $\log h \in L^1(\partial\mathbb{D})$

(β) there exists an $f \in N$, $f \not\equiv 0$ such that

$$|f(\zeta)| = \lim_{r\to 1-0} |f(r\zeta)| = h(\zeta) \qquad \text{a.e. } \zeta \in \partial\mathbb{D} \tag{3}$$

We also have the equivalent statements (α_p) and (β_p):

(α_p) $\log h \in L^1(\partial\mathbb{D})$, $h \in L^p(\partial\mathbb{D})$

(β_p) there exist an $f \in H^p$, $f \not\equiv 0$ such that (3) holds.

The inclusion $h \in L^p(\partial\mathbb{D})$ implies $\int_{\partial\mathbb{D}} \log h \, d\theta < \infty$, hence the statement ($\alpha_p$) may be rewritten in the form

(α'_p) $\int_{\partial\mathbb{D}} \log h \, d\theta > -\infty$, $h \in L^p(\partial\mathbb{D})$.

Therefore the equivalence of (α'_p) and (β_p) yields a uniqueness theorem useful in applications,

$$f\in H^p, \int_{\partial\mathbb{D}} \log|f|\,d\theta=-\infty \Rightarrow f\equiv 0 \tag{4}$$

III. Nevanlinna's theorem also contains a full description of the zero-sets of functions N : if $\{\alpha_k\}$ is a countable set in $\mathbb{D}$ then the following are equivalent:

(γ) there exists a function $f\not\equiv 0$, $f\in N$, such that $f^{-1}(0)=\{\alpha_k\}$

(δ) $\sum(1-|\alpha_k|)<\infty$

IV. Nevanlinna's factorization is multiplicative:

$$e^{(fg)}=e^f\cdot e^g, \quad I_{fg}=I_f\cdot I_g .$$

These relations form an analytic basis of many important theorems concerning the structure of ideals or invariant subspaces in some spaces of analytic functions. A well-known (but not the only one) example is given by the famous Beurling theorem on the shift operator $f\longmapsto zf$ on H^2.

4. We are now able to state (in a general form) four problems which are treated in the present notes.

I. What are the classes $X\subset\mathfrak{D}$ which are invariant with respect to the "isolation of the zeros" ?

II. What are the moduli $|f|_{\partial\mathbb{D}}$ of a given class $X\subset\mathfrak{D}'$?

III. What are the zero sets of functions of a given class X ?

IV. What is the structure of closed ideals X if X is a Banach algebra (or what is the structure of shift invariant subspaces if X is a Banach space) ?

We postpone a detailed discussion and now only stress that we are going to study usual smooth analytic functions.

DEFINITION. Following V.P.Havin [4] we say that a class $X\subset\mathfrak{D}$

possesses the $(\mathfrak{F})$-property if for any $f\in X$ and for any inner function I dividing I_f the function fI^{-1} belongs to X.

We have already seen that H^p and $\mathfrak{D}$ possess the $(\mathfrak{F})$-property. That is a simple consequence of the factorization theorem. Because $f/I\in\mathfrak{D}$ and $|f/I|/\partial\mathbb{D}=|f|_{\partial\mathbb{D}}$ a.e. and the classes H^p are defined only in terms of $|f|_{\partial\mathbb{D}}$. But the statement that the disc-algebra C_A i.e. $f\in\mathfrak{D}$ (f is continuous in $\bar{\mathbb{D}}$) possesses the $(\mathfrak{F})$-property is deeper (this result was first stated by W.Rudin [5] in connection with his investigation of closed ideals in C_A). The disc-algebra C_A in contradiction to H^p and $\mathfrak{D}$ contains inner functions only as exception. Hence the $(\mathfrak{F})$-property in C_A is due to specific interference of outer and inner factors.

Much deeper than in C_A is an unexpected result of L.Carleson [3] , who has discovered the $(\mathfrak{F})$-property in the class

$$W^2_{1A}=\{f\in\mathfrak{D}:\iint|f'(x+iy)|^2dx\,dy<\infty\}.$$

Moreover L.Carleson has succeeded in describing all of the parameters $\mu(f),\{\alpha_K\},|f|_{\partial\mathbb{D}}$ of the factorization (2). Functions in W^2_{1A} (analytic functions with finite Dirichlet integral) have appropriate smoothness on $\partial\mathbb{D}$ and that numbers the investigation of the interplay of outer and inner factors.

B.I.Korenblum [6] , [7] has shown that the classes $H^2_n=\{f: f^{(n)}\in H^2\}$ possess the (F)-property. Such functions are already really smooth up to the boundary. Using a development of the method of [6] , [7] . V.P.Havin [4] has proved the same in the classes

$$H^p_n=\{f\in H^p: f^{(n)}\in H^p\},\quad n\geq 1,\quad 1<p<\infty$$

and

$$\Lambda^{n+\alpha}=\{f: f^{(n)}\in\Lambda^{\alpha}\}$$

$$\Lambda^\alpha=\{f: |f(z)-f(\zeta)|\le C_f|z-\zeta|^\alpha,\ z,\zeta\in\mathbb{D}\},\ 0<\alpha<1.$$

Independently and at the some time the (F)-property for $\Lambda^{n+\alpha}$ and some other classes was stated by F.A. Shamoyan.

The method mentioned above in some situations permits one to avoid an ingenious analysis of outer and inner factors. The main ideas are the following.

We define a Toeplitz operator $T_{\bar a}$:

$$(T_{\bar a} f)_{(\tau)}=\frac{1}{2\pi i}\int\limits_{\partial\mathbb{D}}\frac{\bar a(\zeta)f(\zeta)}{\zeta-\tau}d\zeta,\quad f\in X,\ a\in H^\infty.$$

If we suppose that

$$T_{\bar a}X\subset X \qquad \text{for any } a\in H^\infty \tag{5}$$

then X possesses the $(\mathfrak{F})$-property. Indeed, if $a=I$, I is an inner function, I divides I_f then $T_{\bar I}f=fI^{-1}$ by the Cauchy formula. Following V.P.Havin [4] we call the property (5) of a class X the (K)-property. In [6a] , [7] , [4] , [41] the $(\mathfrak{F})$-property follows from the (K)-property of the corresponding class. The same implication was obtained in [39] by E.M.Dyn'kin in a different way. J.P.Kahane [46] has applied the best polynomial appronimation and has obtained the (F)-property in Λ^α, $0<\alpha<1$.

Taking in consideration all these results it may look quite natural that all "natural" classes possess the $(\mathfrak{F})$-property.

However, it turned out that the most natural class

$$C_A^n=\{f\in C_A: f^{(n)}\in C_A\}$$

is the most difficult one for the proof of the (F)-property.

It is not hard to prove that these classes do not possess the (K)-property.So it seems that the proof of the (F)-property must rest on a careful analysis of I_f and e^f .

The first paper, in which the $(\mathfrak{F})$-property was studied directly (without use of Toeplitz operators) was that by S.A.Vinogradov and the author [49] (excluding the pioneer work of L.Carleson [3] where studying of the (F)-property was not the main purpose). It was shown in [49] that the space

$$H_1^1=\{f\in C_A : f'\in H^1\}$$

possesses the (F)-property and the space

$$H_1^\infty=\{f\in C_A : f'\in H^\infty\}$$

"almost possesses" the (F)-property. For C_A^n and $H_n^\infty=\{f: f^{(n)}\in \in H^\infty\}$ the problem discussed was solved by the author [51], [53] with the help of a new method which permitted one to study in detail the rate of vanishing of $|f|$ in the vicinity of the critical set $\operatorname{spec} I = B^{-1}(0)\cup \operatorname{supp}\mu_s$, $I=BS$, I divides I_f. In the present volume we apply that method to classes of analytic functions with "varying boundary smoothness" (Ch. 1). The main result of § 1 is the following.

The class $\Lambda_\omega^n(\Phi)$ is a natural generalization of Λ_ω^n and is defined as follows:

$$\Lambda_\omega^n(\Phi)=\{f\in C_A : |f^{(n)}(z)-f^{(n)}(\zeta)|\leqslant$$
$$\leqslant c_\omega\omega(|\Phi((1-\frac{|z-\zeta|}{2})z)|\cdot|z-\zeta|),\quad z,\zeta\in\bar{\mathbb{D}}\}.$$

Φ is an outer function in $\mathbb{D}$ such that $\|\Phi\|_{\partial\mathbb{D}}\in A_1$, A_1 is the Muckenhoupt class, i.e. the class of nonnegative weights which for any are $I\subset\partial\mathbb{D}$ satisfy

$$\int_I|\Phi|\leqslant c|I|\operatorname{ess\,inf}_I|\Phi|.$$

THEOREM 1. Let $n\geqslant 0$, ω be an arbitrary modulus of continuity and $|\Phi|_{\partial\mathbb{D}}\in A_1$. Then the class $\Lambda_\omega^n(\Phi)$ possesses the (F)-property.

It is natural to ask whether the multiplication of a function by its own inner factor retains the function in the considered class of smooth analytic functions as the division does.

If $f \in H^p$ and I is an inner function then obviously $fI \in H^p$. If $f \in C_A$, then fI in general does not belong to C_A but if we know in addition that I divides I_f then it is not difficult to check that $fI \in C_A$. The latter means the following. If I is an inner function then the conditions $f/I \in C_A$ and $fI \in C_A$ are equivalent. However this situation does not occur in classes of smooth analytic functions what is shown by Theorems 2 and 3.

THEOREM 2. Let $f \in \Lambda^n_\omega(\Phi)$, Φ be as in Theorem 1, I be an inner function, $f/I \in C_A$. Suppose that the multiplicity of the zeros of f at the points $\alpha \in \operatorname{spec} I \cap \mathbb{D}$ is at least $n+1$. Then $fI \in \Lambda^n_\omega(\Phi)$.

THEOREM 3. For any modulus of continuity ω there exists a function $f \in A^\infty$, $A^\infty = \bigcap_{n=1}^{\infty} C^n_A$ and a Blaschke product B such that $f/B \in A^\infty$, $fB \notin \Lambda^1_\omega$.

In spite of the abundance of examples of spaces of analytic functions with $(\mathcal{F})$-property, this property is not universal.

The first example of a space without $(\mathcal{F})$-property was pointed out by V.P.Gyrarii [9], who proved that the $(\mathcal{F})$-property is violated in

$$\ell^1_A = \{f \in C_A : \sum_{n \geq 0} |\hat{f}(n)| < \infty\}.$$

Later other examples were discovered:

$$\ell^p_A = \{f \in C_A : \sum_{n \geq 0} |\hat{f}(n)|^p < \infty\}$$

(for $p \in (1, \frac{4}{3})$ see [50] and for $p \in [\frac{4}{3}, 2]$ see [10]),

$$B_0 = \{f \in H^\infty : |f'(z)| = o((1-|z|)^{-1})\}.$$

(J.M.Anderson [11]).

In § 3 of Ch. 1 we exhibit new examples of classes without $(\mathcal{F})$ - property.

THEOREM 4. Let $\{b_n\}$ be any sequence satisfying $1 \le b_n \le c_1 n^{c_2}$. Then the class

$$\{f \in H^1 : \sum_{n \ge 0} b_n |\hat{f}(n)|^p < \infty\}$$

does not possess the $(\mathcal{F})$ -property for $p \in [1, \infty]$, $p \ne 2$.

II . Boundary values of the moduli of smooth analytic functions. Suppose first that the boundary values of a function $f \in \mathcal{D}$ have some smoothness (for example $f \in \mathrm{Lip}\,\alpha$). What then can be said about $|f|\,|_{\partial\mathbb{D}}$? It is clear that $\log|f|$ must be summable on $\partial\mathbb{D}$ and that $|f|\,|_{\partial\mathbb{D}} \in \mathrm{Lip}\,\alpha$ $0 \le \alpha < 1$. That is the only thing which is seen at the first glance But there exist much deeper observations. In [12] V.P.Havin and F.A.Shamoyan proved that for a nonnegative function h, $h \in \mathrm{Lip}\,\alpha$, $0 < \alpha < 1$, $\log h \in L^1(\partial\mathbb{D})$, the outer function f with $|f|\,|_{\partial\mathbb{D}} = h$ satisfies $f \in \Lambda^{\alpha/2}$. This result cannot be improved. There is a function $h \in \mathrm{Lip}\,\alpha$ such that for the corresponding outer function f we have $f \in \Lambda^{\alpha/2+\varepsilon}$ for every $\varepsilon > 0$ (it is mentioned in [13] that a close result is contained in an unpublished paper by Jacobs). Later the theorem mentioned was generalized by V.P. Havin [13] to the class $\mathrm{Lip}\,\omega$ for an arbitrary ω . The results of [12] and [13] pointed out that the boundary smoothness of an outer function must be half that of its modulus, whatever the understanding of the word "smoothness". In connection with his research in Approximation Theory J.Brennan [14] was forced to prove the theorem discussed for $h \in \mathrm{Lip}\,\alpha$, $0 < \alpha < 2$. It is also worth mentioning that the implication $h \in \mathrm{Lip}\,\alpha \Rightarrow {}_e h \in \Lambda^{\alpha/2}$ for any $\alpha > 0$ was used without proof as a crucial tool in papers by Taylor and Williams [15] and by Bruna and Ortega [47] . These authors referred to an unpublished paper by Carleson and Jacobs.

On the other hand in [51] a necessary and sufficient condition for the inclusion ${}_e h \in \Lambda^\alpha$ was found under assumption $h \in Lip\alpha$. However, the form that condition was stated was not convenient for further generalizations.

In Ch. 2 we state a general result in that direction which concerns the scales Λ^α, α is not integer, H^p_n, $1<p<\infty$, $n \geqslant 1$ and $\Lambda^{n-1}Z$, Z is Zygmund class. Theorems 5, 6 and 7 correspond the classes cited. There is one idea of the description which can be realized in different ways depending of the situation. We shall use the common notation

$$L^p_n = \{f : \left(\frac{d}{d\theta}\right)^n f(e^{i\theta}) \in L^p(\partial\mathbb{D})\},$$

$Lip\alpha$ for α not integer and

$$C^{n-1}Z = \{f : \left(\frac{d}{d\theta}\right)^{n-1} f(e^{i\theta}) \in Z(\partial\mathbb{D})\}.$$

We also introduce a specific notation: for a continuous function we put

$$M_h(z) = \max_{\substack{\zeta\in\partial\mathbb{D} \\ |\zeta - \frac{z}{|z|}| \leqslant 1-|z|}} |h(\zeta)|, \quad z\in\mathbb{D}, |z| \geqslant \frac{1}{2}.$$

Now we present a generalized statement of Theorems 5 - 7.

Let X be Λ^α, H^p_n or $\Lambda^{n-1}Z$ and Y be respectively $Lip\alpha$, L^p_n or $C^{n-1}Z$. We put

$$H(X,\varphi,z) = \begin{cases} (1-|z|)^\alpha & \text{if } X=\Lambda^\alpha \\ (1-|z|)^n & \text{if } X=\Lambda^{n-1}Z \\ \varphi\left(\frac{z}{|z|}\right)(1-|z|^n) & \text{if } X=H^p_n. \end{cases}$$

(We emphasize that $H(X,\varphi,z)$ really depends on φ only in the case $X=H^p_n$)

a) Suppose that $f\in X$, $f\not\equiv 0$. For the suitable choice of φ we have the inequality

$$\int_{\partial\mathbb{D}} \left|\log\left|\frac{M_f(z)}{f(\zeta)}\right|\right| \frac{1-|z|^2}{|\zeta-z|^2}\,|d\zeta| \leqslant C \tag{6}$$

which holds at every point $z\in\mathbb{D}\setminus\{0\}$ such that

$$M_f(z) \geqslant H(X,\varphi,z) \tag{7}$$

(C does not depend on z)

b) Suppose that $f\in Y$ and suppose that (6) holds with suitable choice of φ at any point $z\in\mathbb{D}\setminus\{0\}$ satisfying (7). Then ${}_e f\in X$

The boundedness of the integral (6) turns out to be quite a useful tool which permits one to investigate the behaviour of an analytic function rather carefully. Some corollaries of Theorems 5 - 7 are collected in Theorem 8.

THEOREM 8. Let $\int_{\partial\mathbb{D}} \log|f|\,d\theta > -\infty$. Then

$$f\in \mathrm{Lip}\,\alpha \Rightarrow_e f\in \Lambda^{\alpha/2} \tag{8}$$

$$f\in L^p_{2n},\ 1<p<\infty \Rightarrow_e f\in H^p_n \tag{9}$$

$$f_1, f_2\in\Lambda^\alpha,\ 0<\alpha<1,\ h(\zeta)=|f_1(\zeta)|+|f_2(\zeta)| \Rightarrow_e h\in\Lambda^\alpha.$$

The implication (8) strengthens the Carleson-Jacobs result cited by Taylor, Williams, Bruna and Orteya because we do not demamd that f be nonnegative (in our case f can be a complex valued function) : F.A. Shamoyan [48] has proved the implication (9) for $f\geqslant 0$. Both (8) and (9) are ε -strict in the natural sense.

III. Local and global properties of zero-sets of smooth analytic functions.

The zero-sets of functions of Λ^{α} were described by L.Carleson [2] . Recall that the Carleson condition for these classes says:

$$\int_{\partial\mathbb{D}} \log dist(z,E)|dz| > -\infty, \quad E = f^{-1}(0) \tag{10}$$

(That condition was first noticed by Beurling in 1939). Taylor and Williams [15] , and independently B.I.Korenblum [16] for any E satisfying (10) have constructed a function $f \in A^{\infty}$ such that $f^{-1}(0) = E$ (notice that the construction of Taylor and Williams did use the Carleson-Jacobs theorem mentioned above). For $f \in \Lambda_{\omega}$, where ω is a modulus of continuity, i.e. for f satisfying

$$|f(z) - f(\zeta)| \leq C_f \, \omega(|z-\zeta|), \quad z,\zeta \in \overline{\mathbb{D}},$$

the Jensen inequality yields

$$\int_{\partial\mathbb{D}} \log \omega(dist(z,E))|dz| > -\infty, \quad E = f^{-1}(0). \tag{11_{ω}}$$

For a special class of moduli ω (ω being a product of exponents of iterated logarithms) J.Stegbuchner [17] , [18] has proved the convers assertion: if E satisfies (11_{ω}) then there exists a function $f \in \Lambda_{\omega}$ such that $f^{-1}(0) = E$. In Ch. 3 we state a more general fact.

THEOREM 9. Let ω be an arbitrary modulus of continuity, I be an inner function with $\operatorname{spec} I \subset E$. Assume that E satisfies (11_{ω}). Then there exists a function $f \in \Lambda_{\omega}$ such that $f^{-1}(0) = E$, $I = I_f$.

Our construction radically differs from that of [2] , [15] - [18]. Further in § 2 of Ch. 3 we study the local behaviour of a function near its boundary zeros. Our main goal here is to find a right generalization of multiplicity in the case of boundary zeros. We propose two ways for that (which are nevertheless, closely intertwined).

First it seems natural to characterize the multiplicity of a boundary zero $\alpha \in \partial \mathbb{D}$ of a function f in terms of a majorant H such that

$$|f(r\alpha)| \leq H(r),$$

$$H(r) \to 0 \quad \text{for } r \to 1-0.$$

Second, to any function $f \in N$ we may attach a local characteristic

$$N_f(\alpha,\delta) = \frac{1}{2\pi} \int_{|\zeta-\alpha|<\delta} \log^+ \frac{1}{|f|} |d\zeta| + \mu_f^+(\{|\zeta-\alpha|<\delta\}) + \sum_{|\alpha_K-\alpha|<\delta} (1-|\alpha_K|^2), \tag{12}$$

where μ_f^+ is the positive part of the singular measure $\{\alpha_K\}$ are the zeros of f in $\mathbb{D}$.

In this notation the multiplicity of α increases for the faster decrease of the majorant H or for the slower decrease of $N_f(\alpha,\delta)$. It is easy to check that any function $f \in N$, $f \not\equiv 0$ cannot decrease faster than $\exp\left(-\frac{C}{1-r}\right)$, for a $C = C(f) > 0$. On the other hand, in case the positive part of the singular measure μ_f contains an atom at α, a majorant $\exp\left(-\frac{C_0}{1-r}\right)$ may occur even for $f \in A^\infty$. Hence the smoothness expressed in terms of the scales Λ^α or H^p_n does not restrict much the multiplicity of a boundary zero.

This question is more essential for classes of functions with rare Taylor spectrum. If we apply the method of H-majorants to the description of multiplicity then our problem turns out to be closely connected with the problem of a possible rate of decreasing of Dirichlet-series with rare exponents along the axes (L.Schwartz [19] , Hirshman and Jenkins [20]).

In Theorem 10 we describe the possible majorants H for functions

$$f(z) = \sum_K a_K z^{n_K} \tag{13}$$

with the property

$$n_k \geq A k^p. \tag{14}$$

For $p>2$ this description is, in a sense, the least one. For $1<p\leq 2$ it is sharper than that of [20] .

Theorem 10 can be applied to any function with rare Taylor coefficients. If we assume in addition that $f\in H^\infty$ then it is natural to expect a moderate behaviour of the characteristic $N_f(\alpha,\delta)$. J.M. Anderson [21] has shown that if f, $f\in H^\infty$, has Hadamard gaps, i.e. $n_{k+1}/n_k \geq q > 1$ (see (13)), then the singular measure of the singular factor of f cannot be "too concentrated". Namely,

$$\varlimsup_{\delta\to 0} \frac{\mu_f(\gamma(\alpha,\delta))}{\delta \log^2 \frac{1}{\delta}} < \infty, \quad \gamma(\alpha,\delta) = \partial\mathbb{D} \cap \{|\zeta-\alpha|<\delta\}.$$

This estimate of J.M.Anderson is a consequence of the Hirshman-Jenkins theorem. In Theorem 11 we obtain a new result of the type discussed. We deal with functions from $\bigcup_{\eta>0} H^\eta$ which satisfy supplementary (14) for $p\geq 2$. We also estimate $N_f(\alpha,\delta)$ and not only $\mu_f(\gamma(\alpha,\delta))$ We prove that

$$\varlimsup_{\delta\to 0} \delta^{\frac{2-p}{p-1}} N_f(\alpha,\delta) < \infty. \tag{15}$$

Moreover, the estimate (15) is precise. We exhibit a function $f\in H^\infty$ such that the left hand side of (15) is positive.

Theorem 12 claims that for such "almost extremal" f the characteristic $N_f(\alpha,\delta)$ behaves rather regularly: if the $\varliminf$ in (15) is positive then the corresponding $\varlimsup$ is finite.

In the last subsection of Ch. 3 we return back to smooth analytic functions. Here we show that the fast decrease to zero of the Taylor coefficients prevents H to be small. This effect can be called an "effect of a compulsory rate of decreasing" of a function with fast decreasing Taylor coefficients. That effect is well illustrated by Theorem 13.

Let $f(z)=\sum_{n\geq 0} a_n z^n$. Suppose that

$$|f(r)|\leq c\exp(-b\log^{\lambda}\frac{1}{1-r}),\quad \lambda>1 \tag{16}$$

and

$$|a_n|\leq c\exp(-\sigma n^{p}),\quad 0<p<\frac{1}{2}. \tag{17}$$

Then the much stronger than (16) estimate holds:

$$|f(r)|\leq c_1\exp(-b_1(1-r)^{-\frac{p}{1-p}}).$$

For $p=\frac{1}{2}$ we obtain the well-known result on quasianalytic functions: if f satisfies (16) and (17) with $p=\frac{1}{2}$ then $f\equiv 0$. It is also clear that the condition $\lambda>1$ in (16) cannot be changed to $\lambda=1$ (for the function $f(z)=1-z$ one has $a_n=0,\ n\geq 2,$ $1-r=\exp(-\log\frac{1}{1-r})$).

IV. Closed ideals in some Banach algebras of analytic functions.

The first result here was obtained by G.E.Shilov [1] who had investigated primary ideals in the disc-algebra C_A . The closed ideals of C_A were described by W.Rudin [5] . In a number of papers closed ideals were studied in algebras ℓ^1_A (V.P.Gurarii [22]), C^n_A, H^2_n, $n\geq 1$ (B.I.Korenblum [23] , [24]), H^p_n, $1<p<\infty$, λ^n_ω for some ω (F.A.Shamoyan [25] - [27]), H^1_n, AB^{α}_{pq} , the analytic Besov classes ([52] , [59]). These results are closely related with the Nevanlinna factorization especially with the (F) -property or some its analogues. The very concept of the (F) -property did appear under the influence of the closed ideals theory.

Despite the variety of analytic tools and methods which have been used in the papers cited above the main result is stated in the standard way. That leads to a concept of "standard ideal" which may be described as follows.

Let X be an algebra such that

$$C_A^{n+1} \subsetneqq X \subset C_A^n$$

for some integer n, $n \geqslant 0$. Let λ be an inner function, $E_0, \dots, E_n$ be closed subsets of $\mathbb{D}$ such that $E_0 \supset E_1 \supset \dots \supset E_n$. Then the ideal

$$\{f \in X : f/\lambda \in C_A^n,\ f^{(\nu)}\big|_{E_k} = 0, \quad 0 \leqslant \nu \leqslant k,\ k = 0, \dots, n\}$$

is called "standard".

In these terms the main result of the papers mentioned can be stated as follows:

All the closed ideals of X are standard.

In Ch. 4 we prove that all the closed ideals of the algebras $X^r_{pq}(\omega, \ell)$ are also standard. That scale contains as subscales the algebras λ^n_ω where ω satisfies

$$\int_0^x \frac{\omega(t)}{t}\, dt + x \int_x^2 \frac{\omega(t)}{t^2}\, dt \leqslant c\,\omega(x), \tag{18}$$

the algebras $AB^{\alpha+r}_{p,q}$, $0 < \alpha \leqslant 1$, $p > \frac{1}{\alpha}$, and the algebras of functions of varying smoothness. The class $X^r_{pq}(\omega, \ell)$ is a generalization of the Besov class and consists of analytic functions such that

$$\left\{ \int_0^1 \frac{dh}{h} \left[\int_{\partial \mathbb{D}} \left| \frac{f^{(r)}(\zeta e^{ih}) - f^{(r)}(\zeta)}{\lambda(\zeta; h)} \right|^p |d\zeta| \right]^{q/p} \right\}^{1/q} < \infty,$$

where $\lambda(\zeta;h)=\omega\left(\int_{\theta}^{\theta+h} \ell(\alpha)d\alpha\right)$, $\zeta=e^{i\theta}$, ω is a modulus of continuity satisfying (18) and ℓ is a positive function such that $\ell^{Q}\in A_1$ for $Q>0$, A_1 is the Muckenhoupt class. If $\omega(t)=t^{\alpha}$, $0<\alpha<1$, $\ell\equiv 1$, then we also assume that $p>\frac{1}{\alpha}$; the similar restrictions are also imposed on p, q, ω, Q in the general case.

The main efforts in §§ 2 - 3 of Ch. 4 are concentrated on the proof of certain approximation assertions in the scope of the well-known Carleman-Beurling-Korenblum method. For that purpose we introduce in $X^{r}_{pq}(\omega,\ell)$ a new equivalent norm. Such a renorming is new even for the Besov classes and is considered in § 1.

NOTATIONS

I. $\mathbb{C}$ - the complex plane

$\mathbb{R}$ - the real line

$\mathbb{D}$ - the unit disc

Π - the upper halfplane

$\mathbb{N}$ - the set of natural numbers

$\mathbb{Z}_+ = \mathbb{N} \cup \{0\}$.

II. Classes of functions on $\partial\mathbb{D}$:

$$C^\alpha,\ 0<\alpha\leq 1: |f(z)-f(\zeta)| \leq c_f|z-\zeta|^\alpha,\ z,\zeta\in\partial\mathbb{D}$$

$$C^{n+\alpha},\ 0<\alpha\leq 1: \left(\frac{d}{dz}\right)^n f \in C^\alpha,$$

$\frac{d}{dz}$ - differentiation along $\partial\mathbb{D}$

$$L^p_n,\ n\geq 0,\ p>0: \left(\frac{d}{dz}\right)^n f \in L^p(\partial\mathbb{D})$$

$$Z: |f(e^{ih}\zeta)-2f(\zeta)+f(e^{-ih}\zeta)| = O(h),\ \zeta\in\partial\mathbb{D}$$

$$C^0 Z \overset{\text{def}}{=} Z,\ C^n Z: \left(\frac{d}{dz}\right)^n f \in Z,\ n\geq 1$$

$$C_\omega: |f(z)-f(\zeta)| = O(\omega|z-\zeta|)),\ z,\zeta\in\partial\mathbb{D} \qquad (1)$$

$$C^n_\omega: \left(\frac{d}{dz}\right)^n f \in C_\omega$$

C^0_ω: change $O(\omega(|z-\zeta|))$ in the right-hand side of (1) to $o(\omega(|z-\zeta|))$

$$\overset{\circ}{C}{}^n_\omega : \left(\frac{d}{dz}\right)^n f \in \overset{\circ}{C}_\omega$$

$$A_1 : h \geqslant 0,$$

$$\int_I h|d\zeta| \leqslant C_h(I) \operatorname*{essinf}_I h \qquad \text{for any arc } I \subset \partial\mathbb{D}.$$

II '. Given an arc I, $I \subset \partial\mathbb{D}$ (or an interval I , $I \subset \mathbb{R}$) and a class X we denote by $X(I)$ the class of restrictions of function in X to I.

III. Classes of analytic function in $\mathbb{D}$.

$\mathcal{A}$ - all analytic functions in $\mathbb{D}$

N - the Nevanlinna class

H^p, $p>0$ - the Hardy class

$C_A : f \in A$, f is continuous in $\overline{\mathbb{D}}$

$$\Lambda^\alpha = C^\alpha \cap C_A$$

$$H^p_n : f^{(n)} \in H^p$$

$$\Lambda^n_\omega = C^n_\omega \cap C_A$$

$$\lambda^n_\omega = \overset{\circ}{C}{}^n \cap C_A$$

$$\Lambda^n z = C^n Z \cap C_A$$

$$\ell^p_A = \{f = \sum_{n\geqslant 0} \hat{f}(n) z^n \in \mathcal{A} : \sum_{n\geqslant 0} |\hat{f}(n)|^p < \infty\}.$$

IV. The Nevanlinna factorization

$\mathcal{B}$ - the set of Blaschken products

S - the set of singular inner functions

$\mathcal{J}$ - the set of inner functions

$\mathcal{J}_b$ - the set of bounded inner functions

if $B \in \mathcal{B}$ then $Z_B = B^{-1}(0)$, $\operatorname{spec} B = \overline{Z}_B$,

if $S \in \mathcal{S}$ then μ_S denotes the measure in the formula

$$S(z) = \exp\left(-\int_{\partial\mathbb{D}} \frac{\zeta+z}{\zeta-z}\, d\mu_S(\zeta)\right), \tag{2}$$

if μ is a singular measure, then $S_\mu \in \mathcal{S}$ denotes the function in (2)

if $\alpha \subset \mathbb{D}$ is a countable set, then

$$B_\alpha(z) = \prod_{\zeta\in\alpha} \frac{\bar{\zeta}}{|\zeta|}\, \frac{\zeta-z}{1-\bar{\zeta}z}$$

if $I = BS \in \mathrm{I}$, then $\operatorname{spec} I = \operatorname{spec} B \cup \operatorname{supp} \mu_S$

if $\log|h| \in L^1(\partial\mathbb{D})$ then

$$e^{h}(z) = \exp\left(\frac{1}{2\pi}\int_{\partial\mathbb{D}} \log|h(\zeta)|\, \frac{\zeta+z}{\zeta-z}\, |d\zeta|\right)$$

if $f \in N$, then $f = e^{f}\cdot I_f = e^{f}\cdot c_f \cdot B_f \cdot S_f$ is the Nevanlinna factorization of f .

For an outer functions in the halfplane Π we use the following notation

$$E^{\varphi}(\zeta) = \exp \frac{1}{\pi i}\int_{\mathbb{R}} \log|\varphi(t)|\, \frac{dt}{t-\zeta}, \quad \zeta \in \Pi.$$

V. Some technical notations

$$D(z,\delta) = \{\zeta \in \mathbb{C} : |\zeta - z| < \delta\}; \; D(z,\delta,\Delta) = D(z,\Delta)\setminus D(z,\gamma)$$

$$H^* f(\zeta) = \sup_{\substack{\partial\mathbb{D}\supset I \ni \zeta \\ \zeta \text{ is the middle of } I}} \left| \int_{\partial\mathbb{D}} \frac{f(t)}{t-\zeta}\, dt \right|, \quad f \in L^1(\partial\mathbb{D}), \; \zeta \in \partial\mathbb{D}$$

$$f_r^*(\zeta) = \sup_{\partial\mathbb{D}\supset I\ni\zeta} \left(\frac{1}{|I|}\int_I |f|^r\, |dz|\right)^{1/r}, \quad 1 \leqslant r < \infty,$$

$$P_r(f; z; \zeta) = \sum_{\nu=0}^{r} \frac{1}{\nu!} \left(\frac{d}{dz}\right)^{\nu} f(z)\cdot(\zeta - z)^{\nu}$$

$a \lesssim b$ denotes $|a| \leq C|b|$, C is an absolute constant

$$a \asymp b : a \lesssim b, \quad \text{and} \quad b \lesssim a$$

$$a \underset{\alpha,\dots,\omega}{\lesssim} b : |a| \leq C(\alpha,\dots,\omega)|b|$$

$$a \underset{\alpha,\dots,\omega}{\asymp} b : a \underset{\alpha,\dots,\omega}{\lesssim} b \quad \text{and} \quad b \underset{\alpha,\dots,\omega}{\lesssim} a$$

$Z_f(\Omega)$ is the zero-set of a function $f \in C(\Omega)$

$$q_n' = 1 - \frac{1}{4(n+1)^2}, \quad q_n'' = 1 + \frac{1}{4(n+1)^2}.$$

VI. The numeration of lemmas and formulas is independent in each chapter.

CHAPTER 1

The (F)-property

In this chapter we enlarge the list of classes of analytic functions possessing the (F) property. The main contribution concerns the class $\Lambda^n_\omega(\Phi)$. Here Φ is an outer function in $\mathbb{D}$ with $|\Phi|\,|_{\partial\mathbb{D}} \in A_1$, A_1 is the Muckenhoupt class, ω is an arbitrary modulus of continuity. We say that $f \in \Lambda^n_\omega(\Phi)$ if $f \in N$ and

$$|f^{(n)}(z_1)-f^{(n)}(z_2)| \leqslant c_f\,\omega(|\Phi(\check{z}_{z_1,z_2})|\cdot|z_1-z_2|),\ z_1,z_2\in\mathbb{D},\ n\geqslant 0, \tag{1}$$

where $\check{z}_{z_1,z_2}$ is the farthest point in

$$\left\{\zeta\in\mathbb{D}: \left|\zeta-\frac{z_1+z_2}{2}\right| = \frac{|z_1-z_2|}{2}\right\} \quad \text{from } \partial\mathbb{D}$$

This result is stated in § 1. In § 2 we consider the multiplication of a function by its own inner factor. We find an effect which distinguishes the classes with smoothness not exceeding 1 from that of higher smoothness. Namely the multiplication by such a factor retains the small smoothness of a function but not necessarily retains a higher smoothness. In § 3 we exibit new examples of classes without (F)-property.

§ 1. The (F)-property for $\Lambda^n_\omega(\Phi)$.

Here we prove theorem 1 on the (F)-property of $\Lambda^n_\omega(\Phi)$ (see the definition above). The proof is divided into a number of steps. The steps 1.0 and 1.1 are purely technical. In 1.2 we examine the influence of the inner factor of a function on its behaviour. In 1.3 we prove the key Lemma 1.11. The proof is similar to that we use in 1.4 to finish the proof of Theorem 1.

We now introduce the class $H_n^\infty(\Phi)$, where Φ is an outer function with $|\Phi\|_{\partial\mathbb{D}}\in A_1$:

$$f\in H_n^\infty(\Phi) \quad \text{if} \quad f\in N,\ |f^{(n)}(z)|\leqslant C_f|\Phi(z)|,\ z\in\mathbb{D},\ n\geqslant 1. \tag{2}$$

THEOREM 1. The classes $\Lambda_\omega^n(\Phi)$ and $H_n^\infty(\Phi)$ possess the (F)-property.

REMARK 1. If $\omega_1(t)\equiv t$ then it is clear that $H_n^\infty(\Phi)=\Lambda_{\omega_1}^{n-1}(\Phi)$. So it is sufficient to prove the result only for $\Lambda_\omega^n(\Phi)$.

REMARK 2. It follows from subsection 1.0 below that $\Lambda_\omega^n(\Phi)\subset C_A$, $n\geqslant 0$ for any ω and any Φ, $|\Phi\|_{\partial\mathbb{D}}\in A_1$.

REMARK 3. Using the arguments of P.M.Tamrazov [28] and well-known properties of the class A_1, we obtain the following result.

THEOREM A. Let $f\in C_A$ and suppose that for any points $z_1,z_2\in\partial\mathbb{D}$

$$|f^{(n)}(z_1)-f^{(n)}(z_2)|\leqslant c\,\omega\left(|\Phi((1-\tfrac{(z_1-z_2)}{2})z_2)|\cdot|z_1-z_2|\right),\quad n\geqslant 0.$$

Then $f\in\Lambda_\omega^n(\Phi)$.

REMARK 4. In the proof of Theorem 1 we in fact obtain a more general result. Namely, there exists a constant $C_{n,\Phi}$ depending only on n and Φ such that if

$$|f^{(n)}(z)-f^{(n)}(\zeta)|\leqslant\omega(|\Phi(\check{z}_{z,\zeta})||z-\zeta|),\quad z,\zeta\in\mathbb{D},$$

where I is an inner function with $f/I\in C_A$, then

$$\left|\left(\frac{f}{I}\right)^{(n)}(z)-\left(\frac{f}{I}\right)^{(n)}(\zeta)\right|\leqslant C_{n,\Phi}\,\omega(|\Phi(\check{z}_{z,\zeta})|\cdot|z-\zeta|),\quad z,\zeta\in\mathbb{D}.$$

1.0. Some technical preparations.

Below we use the notation $\check{z}_{z_1,z_2}$ introduced in Theorem 1.

LEMMA 1.1. Let $z_1, z_2 \in \partial\mathbb{D}$,

$\zeta_1, \zeta_2 \in \mathbb{D}\left(\frac{z_1+z_2}{2}, \frac{|z_2-z_1|}{2}\right)$, $|\Phi|\big|_{\partial\mathbb{D}} \in A_1$. Then $|\Phi(\check{z}_{z_1,z_2})| \underset{\Phi}{\lesssim} |\Phi(\check{z}_{\zeta_1,\zeta_2})|$

PROOF. In accordance with [29] for $\check{z} = \check{z}_{z_1,z_2}$, $\check{\zeta} = \check{\zeta}_{\zeta_1,\zeta_2}$ we have

$$|\Phi(\check{z})| \asymp \frac{1}{|\gamma(\check{z})|}\int_{\gamma(\check{z})} |\Phi(t)||dt|,\ |\Phi(\check{\zeta})| \asymp \frac{1}{|\gamma(\check{\zeta})|}\int_{\gamma(\check{\zeta})} |\Phi(t)||dt|,$$

where

$$\gamma(w) = \{\tau \in \partial\mathbb{D} : |\arg \tfrac{\tau}{w}| \leqslant 2 \arcsin(1-|w|\},\quad w \in \mathbb{D}.$$

Next, for the arc $\gamma^*, |\gamma^*| = 3|\gamma(\check{z})|$ centered at the middle of γ , the standard arguments [30] yield

$$\frac{1}{|\gamma(\check{z})|}\int_{\gamma(\check{z})} |\Phi(t)||dt| \asymp \operatorname*{ess\,inf}_{\gamma(\check{z})} |\Phi| \asymp \operatorname*{ess\,inf}_{\gamma^*} |\Phi|,$$

$$\frac{1}{|\gamma(\check{\zeta})|}\int_{\gamma(\check{\zeta})} |\Phi(t)||dt| \asymp \operatorname*{ess\,inf}_{\gamma(\check{\zeta})} |\Phi| \asymp \operatorname*{ess\,inf}_{\gamma^*} |\Phi|,$$

since $\gamma(\check{\zeta}) \subset \gamma^*$. ●

LEMMA 1.2. [30] . For some q, $0 < q < 1$ we have

$$|\Phi(z)| \underset{\Phi}{\lesssim} (1-|z|)^{-q}. \quad ●$$

LEMMA 1.3. Let $z_1, z_2 \in \partial\mathbb{D}$, $|z_1 - z_2| = \sigma$,

$\zeta_1, \zeta_2 \in \mathbb{D}\left(\frac{z_1+z_2}{2}, \frac{|z_1-z_2|}{2}\right)$, ω be an arbitrary modulus of continuity. Then

$$\frac{|\zeta_1-\zeta_2|}{\sigma}\,\omega\big(2|\Phi(\check{z}_{z_1,z_2})|\sigma\big) \lesssim \big(|\Phi(\check{z}_{\zeta_1,\zeta_2})|\cdot|\zeta_1-\zeta_2|\big).$$

PROOF. The definition of modulus of continuity and Lemma 1.1 yield

$$\frac{|\zeta_1-\zeta_2|}{\sigma}\,\omega\big(2|\Phi(\check{z}_{z_1,z_2})|\sigma\big) \leqslant 2\omega\big(|\Phi(\check{z}_{z_1,z_2})|\cdot|\zeta_1-\zeta_2|\big) \underset{\Phi}{\lesssim} \omega\big(|\Phi(\check{z}_{\zeta_1,\zeta_2})|\cdot|\zeta_1-\zeta_2|\big). \quad ●$$

1.1. The crucial lemmas.

In what follow the sign $\preccurlyeq$ means $\underset{\Phi,n}{\leqslant}$.

LEMMA 1.4. Let Ω, $\Omega \subset \overline{D}$ be a convex set, $\operatorname{diam}\Omega \leqslant \rho$, not reducing to a segment and let f be an analytic function in Ω and

$$|f^{(n)}(z) - f^{(n)}(\zeta)| \leqslant \omega(|\Phi(\check{z}_{z,\zeta})|\cdot|z-\zeta|), \quad z,\zeta \in \Omega, \quad n \geqslant 0. \tag{1.1.0}$$

suppose further that f has $n+1$ zeros in $\overline{\Omega}$ (taking account of multiplicities) then

$$|f^{(\nu)}(z)| \leqslant C_{n\Phi 1}\, \rho^{n-\nu}\, \omega(|\Phi(\check{z}_{\Omega})|_{\rho}), \quad z \in \overline{\Omega}, \quad \nu = 0,1,\dots,n, \tag{1.1.1}$$

where $\overline{z}_{\Omega} = \check{z}_{z_1,z_2}$, $|z_1, z_2|$ is a diameter of Ω.

PROOF. We use the induction in n.

If $n=0$ and $f(a)=0$, then

$$|f(z)| = |f(z) - f(a)| \leqslant \omega(|\Phi(\check{z}_{z,a})||z-a|) \leqslant \omega(|\Phi(\check{z}_{\Omega})|\rho).$$

We consider now the function f for which (1.1.0) is satisfied for some $n>0$. Let $a \in \Omega$, $f(a)=0$, $\varphi(z) = f(z)/(z-a)$. The function φ has n zeros in $\overline{\Omega}$ and

$$\varphi(z) = \sum_{\nu=1}^{n} \frac{1}{\nu!} f^{(\nu)}(a)(z-a)^{\nu-1} + \frac{1}{(n-1)!}\frac{1}{z-a}\int_a^z (z-t)^{n-1}(f^{(n)}(t) - f^{(n)}(a))\,dt,$$

$$\varphi^{(n-1)}(z_1) - \varphi^{(n-1)}(z_2) = \sum_{\nu=0}^{n-1} C_{n-1}^{\nu} \frac{(-1)^{\nu}}{(z_1-a)^{\nu+1}} \int_a^{z_1} (z_1-t)^{\nu}(f^{(n)}(t) - f^{(n)}(a))\,dt -$$

$$- \sum_{\nu=0}^{n-1} C_{n-1}^{\nu} \frac{(-1)^{\nu}}{(z_2-a)^{\nu+1}} \int_a^{z_2} (z_2-t)^{\nu}(f^{(n)}(t) - f^{(n)}(a))\,dt \overset{def}{=\!=}$$

$$\overset{def}{=\!=} \sum_{\nu=0}^{n-1} (-1)^{\nu} C_{n-1}^{\nu} (S_{\nu}^{(1)} - S_{\nu}^{(2)}); \tag{1.1.2}$$

$$\varphi^{(n)}(z)=\frac{f^{(n)}(z)-f^{(n)}(a)}{z-a}+\sum_{\nu=0}^{n}\nu C_n^{\nu}\frac{(-1)^{\nu}}{(z-a)^{\nu+1}}\int_a^z(z-t)^{\nu-1}(f^{(n)}(t)-f^{(n)}(a))dt. \qquad (1.1.3')$$

Let $|z_2-a|\leqslant|z_1-a|=\Delta$.

CASE 1. $|z_2-a|\leqslant\frac{1}{2}\Delta$. Then $|z_2-z_1|\geqslant\frac{1}{2}\Delta$ and Lemma 1.3 yields

$$|S_{\nu}^{(j)}|=\left|\frac{1}{(z_j-a)^{\nu+1}}\int_a^{z_j}(z_j-t)^{\nu}(f^{(n)}(t)-f^{(n)}(a))dt\right|\leqslant$$

$$\leqslant\frac{|z_j-a|^{\nu+1}}{|z_j-a|^{\nu+1}}\,\omega(|\Phi(\check{z}_{z_j,a})||z_j-a|\leqslant\omega(|\Phi(\check{z}_{z_1,z_2})||z_1-z_2|), \qquad (1.1.3)$$

From (1.1.2) and (1.1.3) we obtain

$$|\varphi^{(n-1)}(z_1)-\varphi^{(n-1)}(z_2)|\leqslant\omega(|\Phi(\check{z}_{z_1,z_2})||z_1-z_2|).$$

CASE 2. $\frac{1}{2}\Delta<|z_2-a|\leqslant\Delta$. Let $\delta=|z_2-z_1|$, $\delta\leqslant 2\Delta$. Then

$$S_{\nu}^{(1)}-S_{\nu}^{(2)}=\left[\frac{1}{(z_1-a)^{\nu+1}}-\frac{1}{(z_2-a)^{\nu+1}}\right]\int_a^{z_2}(z_1-t)^{\nu}(f^{(n)}(t)-f^{(n)}(a))dt+$$

$$+\frac{1}{(z_2-a)^{\nu+1}}\left[\int_a^{z_1}(z_1-t)^{\nu}(f^{(n)}(t)-f^{(n)}(a))dt-\int_a^{z_2}(z_2-t)(f^{(n)}(t)-f^{(n)}(a))dt=\right.$$

$$=\left[\frac{1}{\cdots}-\frac{1}{\cdots}\right]\int_a^{z_1}\cdots+\frac{1}{(z_2-a)^{\nu+1}}\left[\int_a^{z_1}((z_1-t)^{\nu}-(z_2-t)^{\nu})(f^{(n)}(t)-f^{(n)}(a))dt+\right.$$

$$\left.+\int_{z_2}^{z_1}(z_2-t)^{\nu}(f^{(n)}(t)-f^{(n)}(a))dt\right]\stackrel{def}{=\!=}r_1+r_2+r_3. \qquad (1.1.4)$$

Next, by Lemma 1.3 we can write again

$$|r_1|\leqslant\frac{(\nu+1)\delta}{(\Delta/2)^{\nu+2}}\Delta^{\nu+1}\,\omega(|\Phi(\check{z}_{z_1,a})|\Delta)\leqslant\omega(|\Phi(\check{z}_{z_1,z_2})|\delta), \qquad (1.1.5)$$

$$|r_2| \leq \frac{\nu}{(\Delta/2)^{\nu+1}}(4\Delta)^{\nu}\delta\,\omega(|\Phi(\check{z}_{z_1,a})|\Delta) \leq \omega(|\Phi(\check{z}_{z_1,z_2})|\delta), \tag{1.1.6}$$

$$|r_3| \leq \frac{\delta^{\nu+1}}{(\Delta/2)^{\nu+1}}\omega(|\Phi(\check{z}_{z_1,z_2})|\Delta) \leq \omega(|\Phi(\check{z}_{z_1,z_2})|\delta). \tag{1.1.7}$$

So from (1.1.3) - (1.1.7) we have

$$|\varphi^{(n-1)}(z_1) - \varphi^{(n-1)}(z_2)| \leq \omega(|\Phi(\check{z}_{z_1,z_2})||z_1 - z_2|), \quad z_1, z_2 \in \bar{\Omega}. \tag{1.1.8}$$

Argueing as in the proof of (1.1.3) - (1.1.7), we obtain

$$|\varphi^{(n)}(z)| \leq \frac{\omega(|\Phi(\check{z}_{z,a})||z-a|)}{|z-a|}. \tag{1.1.9}$$

Now the induction and (1.1.8) yield

$$|\varphi^{(\nu)}(z)| \leq \rho^{n-\nu-1}\omega(|\Phi(\check{z}_{\Omega})|\rho), \qquad \nu = 0,\dots,n-1. \tag{1.1.10}$$

Finally, we have

$$f(z) = (z-a)\varphi(z), \quad f^{(\nu)}(z) = (z-a)\varphi^{(\nu)}(z) + \nu\varphi^{(\nu-1)}(z), \ \nu > 0,$$

and then (1.1.10) implies for $\nu = 0$ that

$$|f(z)| \leq \rho|\varphi(z)| \leq \rho^{n}\omega(|\Phi(\check{z}_{\Omega})|\rho).$$

For $1 \leq \nu \leq n-1$ (1.1.10) again yields

$$|f^{(\nu)}(z)| \leq \rho \cdot \rho^{n-\nu-1}\omega(|\Phi(\check{z}_{\Omega})|\rho) = \rho^{n-\nu}\omega(|\Phi(\check{z}_{\Omega})|\rho).$$

For $\nu = n$ (1.1.9) and (1.1.10) yield

$$|f^{(n)}(z)| \leq |z-a| \cdot \frac{\omega(|\Phi(\check{z}_{z,a})||z-a|)}{|z-a|} + \omega(|\Phi(\check{z}_{\Omega})|\rho) \leq \omega(|\Phi(\check{z}_{\Omega})|\rho).$$

q.e.d. ●

LEMMA 1.5. a) Let $f \in \Lambda^{n}_{\omega}(\Phi)$,

$$|f^{(n)}(z_1)-f^{(n)}(z_2)| \leqslant \omega(|\Phi(\check{z}_{z_1,z_2})||z_1-z_2|), \quad z_1,z_2\in\mathbb{D}, \tag{1.1.11}$$

$$f(a)=0, \qquad a\in\mathbb{D}.$$

Then for $f_0(z) \overset{\text{def}}{=\!=} \dfrac{1-\bar{a}z}{z-a} f(z)$ we have

$$|f^{(n)}(z_1)-f^{(n)}(z_2)| \lesssim \omega(|\Phi(\check{z}_{z_1,z_2})||z_1-z_2|), \quad z_1,z_2\in\mathbb{D}; \tag{1.1.12}$$

b) Let $\Omega=\{\zeta: |\zeta|<1, |\zeta-z_0|<h\}$, $z_0\in\partial\mathbb{D}$, $0<h<1$, $f\in C_A(\bar{\Omega})$.

Suppose that for f (1.1.11) is satisfied for any $z_1,z_2\in\Omega$. Let $a\in\Omega$, $|a-z_0|<qh$, $q<1$. Then for f_0 (1.1.12) is valid for any $z_1,z_2\in\Omega$, where the sign $\lesssim$ means $\underset{\Phi,n,q}{\prec}$.

The proof is similar to that of Lemma 1.4. ●

1.2. The influence of the inner factor.

LEMMA 1.6. Let $z\in\partial\mathbb{D}$, I be an inner function, $|I|\leqslant 1$,

$d=\{\zeta:|\zeta|<1, |\zeta-z_0|<h\}$. Then for $|\zeta-z|<\frac{1}{2}\Delta$

we have

$$e^{-c_1(1-|\zeta|)a} \leqslant |I(\zeta)| \leqslant e^{-c_2(1-|\zeta|)a}, \quad |\zeta|\leqslant 1,$$

$$e^{c_2(|\zeta|-1)a} \leqslant |I(\zeta)| \leqslant e^{c_1(|\zeta|-1)a}, \quad |\zeta|\geqslant 1,$$

c_1, c_2, $0<c_2\leqslant c_1<\infty$ are constants.

PROOF. See [51] . ●

LEMMA 1.7. Let $z\in\partial\mathbb{D}$, I be an inner function, $|I|\leqslant 1$,

$$d=\mathrm{dist}(z,\mathrm{spec}\, I)>0, \quad \sigma=\frac{1}{4}\min(d, 1/|I'(z)|).$$

Then for $|\zeta-z|\leqslant\sigma$, $\nu=1,2,\ldots$ we have

$$|I^{(\nu)}(\zeta)| \leq c_\nu \sigma^{-\nu},$$

$$|(1/I)^{(\nu)}(\zeta)| \leq c_\nu \sigma^{-\nu}.$$

PROOF. See [11] . ●

LEMMA 1.8. Let $f \in \Lambda^n_\omega(\Phi)$,

$$|f^{(n)}(z_1) - f^{(n)}(z_2)| \leq \omega(|\Phi(\check{z}_{z_1,z_2})||z_1 - z_2|), \quad z_1, z_2 \in \mathbb{D}, \ n \geq 1;$$

$$0 < \sigma < \frac{1}{200n}, \quad 100n \leq A < \frac{1}{2\sigma}, \ \rho = A\sigma; \ z \in \partial\mathbb{D}, \ z_0 \overset{\text{def}}{=\!=} (1-\rho)z;$$

$$\gamma \overset{\text{def}}{=\!=} \{\zeta : |\zeta - z_0| = \sigma\}, \quad \tau \overset{\text{def}}{=\!=} \{\zeta \in \partial\mathbb{D} : |\zeta - z| \leq 2\sigma\}.$$

Let

$$x = x_\tau \overset{\text{def}}{=\!=} \max_{\zeta \in \tau} |f(\zeta)|,$$

$$y = y_\gamma \overset{\text{def}}{=\!=} \max_{\zeta \in \gamma} |f(\zeta)|.$$

Then

$$x \leq 2A^n y + C_{\Phi,n}\rho^n \omega(|\Phi(z_0)|\rho), \tag{1.2.1}$$

$$|f(\zeta)| \leq x + 12n!\,A^n y \left(\frac{|\zeta - z| + \sigma}{\sigma}\right)^n + C_{\Phi,n}(|\zeta - z| + \rho)^{n+1} \frac{\omega(|\Phi(z_0)|\rho)}{\rho}, \tag{1.2.2}$$

$$\zeta \in \partial\mathbb{D}.$$

Proof. The Cauchy-inequalities yield

$$|f^{(\nu)}(z_0)| \leq \frac{\nu!}{\sigma^\nu} y, \qquad \nu = 1, \ldots, n.$$

Next, for the point $\zeta_0 \in \tau$ satisfying $|f(\zeta_0)| = x$ we have

$$f(\zeta_0)=f(z_0)+\sum_{\nu=1}^{n}\frac{f^{(\nu)}(z_0)}{\nu!}(\zeta_0-z_0)^{\nu}+\frac{1}{(n-1)!}\int_{z_0}^{\zeta_0}(\zeta_0-t)^{n-1}(f^{(n)}(t)-f^{(n)}(z_0))\,dt,$$

$$x\leq y+\sum_{\nu=1}^{n}(A+2)^{\nu}y+C_{\Phi,n}\rho^{n}\omega(|\Phi(z_0)|\rho)\leq$$

$$\leq 2A^{n}y+C_{\Phi,n}\rho^{n}\omega(|\Phi(z_0)|\rho)$$

since $A\geqslant 100n$. The Taylor formula and Lemma 1.3 imply

$$|f^{(\nu)}(z)|\leqslant 4n!\frac{A^{n-\nu}}{\sigma^{\nu}}y+C_{\Phi,n}\rho^{n-\nu}\omega(|\Phi(z_0)|\rho),\quad 0\leqslant\nu\leqslant n. \tag{1.2.3}$$

Now (1.2.1) and (1.2.2) are consequences of the Taylor formula, (1.2.3) and Lemma 1.3 ●

We preserve the notation of Lemma 1.8 in the following lemma.

LEMMA 1.9. Let $z^{\circ}\in\gamma$, and U be an arbitrary number. Then

$$\frac{1}{2\pi}\int_{\partial\mathbb{D}}\log\Big[x+12n!\,A^{n}y\Big(\frac{|\zeta-z|+\sigma}{\sigma}\Big)^{n}+U(|\zeta-z|+\rho)^{n+1}\Big]\frac{1-|z^{\circ}|^{2}}{|\zeta-z^{\circ}|^{2}}|d\zeta|\leqslant$$

$$\leqslant\log\big[C_{n0}\cdot 12n!\,2^{2n+20}(x+A^{2n+1}y+U\rho^{n+1})\big]$$

for any x,y,A,U,ρ,z.

PROOF. See Lemma 5 in [53] . ●

LEMMA 1.10. We take C_2 from Lemma 1.6, C_{n0} from Lemma 1.9 and put

$$C_n=C_{n0}\cdot 12n!\cdot 2^{2n+20}$$

where A denotes the root of the equation

$$C_nA^{2n+1}e^{-\frac{1}{2}C_2A}=\frac{1}{8}.$$

Suppose that $f\in\Lambda_{\omega}^{n}(\Phi)$, $n\geqslant 1$,

$$|f^{(n)}(z_1)-f^{(n)}(z_2)|\leqslant\omega(|\Phi(\check{z}_{z_1,z_2})||z_1-z_2|),\quad z_1,z_2\in\mathbb{D},$$

$z\in\partial\mathbb{D}$, I is an inner function, $|I|\leqslant 1$, $f/I\in C_A$,

$$\ell = \mathrm{dist}(z, \operatorname{spec} I) > 0, \quad a \overset{\mathrm{def}}{=} |I'(z)| > \frac{4A}{\ell}.$$

Then for $\zeta \in \mathbb{D}$, $|\zeta - z| \leqslant \sigma \overset{\mathrm{def}}{=} 1/a$ we have

$$|f^{(\nu)}(\zeta)| \leqslant C_n \sigma^{n-\nu} \omega(|\Phi((1-\sigma)z)|\sigma), \quad \nu = 0, 1, \dots, n$$

PROOF. We put $\rho = A\sigma$, $z_0 = (1-\rho)z$, and with the points z, z_0 the numbers σ, ρ, A (notice that $A\sigma < \frac{1}{4}\ell$), and the function f, we relate an arc τ, a circle τ, and the numbers $x = x_\tau$, $y = y_\gamma$ as in Lemma 1.8. Let further $z^\circ \in \gamma$ be such that $|f(z^\circ)| = y$. Let F be an outer factor of f, $\zeta_1 = z^\circ/|z^\circ|$. From Lemma 1.6 we obtain

$$y \leqslant |F(z^\circ)||I(z^\circ)| \leqslant |F(z^\circ)| \leqslant |F(z^\circ)| e^{-c_2 a(1-|z^\circ|)} \leqslant e^{-\frac{Ac_2}{2}} |F(z^\circ)|. \tag{1.2.5}$$

Lemmas 1.8 and 1.9 imply

$$\log|F(z^\circ)| = \frac{1}{2\pi} \int_{\partial\mathbb{D}} \log|f(\zeta)| \frac{1-|z^\circ|^2}{|\zeta - z^\circ|^2} |d\zeta| \leqslant$$

$$\leqslant \frac{1}{2\pi} \int_{\partial\mathbb{D}} \frac{1-|z^\circ|^2}{|\zeta - z^\circ|^2} \log\Big[x + 12n!\, A^n y \Big(\frac{|\zeta - z| + \sigma}{\sigma}\Big)^n +$$

$$+ C_{n,\Phi}(|\zeta - z| + \rho)^{n+1} \frac{\omega(|\Phi(z_0)|\rho)}{\rho}\Big] |d\zeta| \leqslant$$

$$\leqslant \log[C_{n0} \cdot 12\, n! \cdot 2^{2n+20} (x + A^{2n+1} y + C_{n,\Phi} \rho^n \omega(|\Phi(z_0)|\rho))], \tag{1.2.6}$$

and (1.2.5) and (1.2.6) yield

$$y \leqslant C_n e^{-\frac{Ac_2}{2}} (x + A^{2n+1} y + C_{n,\Phi} \rho^n \omega(|\Phi(z_0)|\rho),$$

$$y \leqslant \frac{1}{8A^{2n+1}} x + \frac{1}{8} y + C_{n,\Phi} \rho^n \omega(|\Phi(z_0)|\rho)),$$

$$y \leq \frac{1}{7A^{2n+1}} x + C_{n,\phi} \rho^n \omega(|\Phi(z_0)|\rho). \tag{1.2.7}$$

Now from (1.2.7) and Lemma 1.8

$$x \leq \frac{2A^n}{7A^{2n+1}} x + C_{n,\phi} \rho^n \omega(|\Phi(z_0)|\rho) \leq \frac{2}{7} x + C_{n,\phi} \cdot \rho^n \omega(|\Phi(z_0)|\rho). \tag{1.2.8}$$

Finally, from (1.2.8) and Lemma 1.3 we obtain

$$x \leq C_{n,\phi} \sigma^n \omega(|\Phi((1-\sigma)z)|\sigma). \tag{1.2.9}$$

The relation (1.2.9) is the required estimate (1.2.4) for $\nu = 0$. Using the Taylor formula Cauchy-inequalities Lemma 1.8 and (1.2.7) we obtain the rest estimates (1.2.4) for $\nu = 1, \dots, n$. ●

DEFINITION. Let $I = BS$ be any inner function which is not a Blaschke product with at most n zeros. We put

$$\nu_{I,n}(\alpha) = \begin{cases} 0, & \alpha \notin \operatorname{spec} I \\ n+1, & \alpha \in \operatorname{spec} I \cap \partial\mathbb{D} \\ \text{the multiplicity of the zeros } \alpha, & \alpha \in \operatorname{spec} I \cap \mathbb{D} \end{cases}$$

$$d_n(z) = d_{n,I}(z) = \inf \Big\{ \delta > 0 : \sum_{\substack{\alpha \in \operatorname{spec} I \\ |\alpha - z| \leq \delta}} \nu_{I,n}(\alpha) \geq n+1 \Big\}.$$

Let $E \subset \mathbb{D}$ be a closed set. Then we denote

$$I|_E = B|_E \cdot S|_E,$$

$$B|_E(z) = \prod_{\alpha \in B^{-1}(0) \cap E} \frac{\bar{z}}{|\alpha|} \cdot \frac{\alpha - z}{1 - \bar{\alpha} z},$$

$$S|_E(z) = \exp\Big(-\int_{\partial\mathbb{D}} \frac{\zeta + z}{\zeta - z} \, d\mu_{S|_E}(\zeta)\Big)$$

$$a_n(z)=a_{n,I}(z)\overset{def}{=}\sum_{\substack{\alpha\in \operatorname{spec} I\\ |\alpha-z|\geq d_{n,I}(z)}}\frac{1-|\alpha|^2}{|1-\bar{\alpha}z|^2}+2\int_{\partial\mathbb{D}\setminus D(z,d_{n,I}(z))}\frac{d\mu_s(\zeta)}{|\zeta-z|^2}.$$

REMARK a) If $f\in\Lambda^n_\omega(\phi)$, $f/I\in C_A$, $z\in \operatorname{spec} I\cap\partial\mathbb{D}$, then z is a zero of f multiplicity $n+1$ [53] p.427.

b) In what follows we assume that the inner function does not coincide with a Blaschke product with no more than n zeros. The reason is that for such simple functions we may use Lemma 1.8.

LEMMA 1.11. Let $f\in\Lambda^n_\omega(\phi)$ and satisfy (1), I be an inner function, $|I|\leq 1$, $f/I\in C_A$, $z\in\partial\mathbb{D}$,

$$d_n(z)=d_{n,I}(z)>0,\qquad a_n(z)=a_{n,I}(z).$$

Then

$$|(f/I)^{(\nu)}(z)|\underset{n,\phi}{\leq}\mu_\nu,\qquad \nu=0,1,\dots,n, \tag{1.2.10}$$

$$|f^{(\nu)}(z)|\underset{n,\phi}{\leq}\mu_\nu,\qquad \nu=0,1,\dots,n, \tag{1.2.11}$$

where

$$\mu_\nu=\min\Big[d_n^{n-\nu}(z)\,\omega(|\phi((1-d_n(z))z)|\,d_n(z)),\ \frac{1}{a_n^{n-\nu}(z)}\,\omega\Big(|\phi((1-\frac{1}{a_n(z)})z)|\,\frac{1}{a_n(z)}\Big)\Big].$$

1.3. Proof of Lemma 1.11.

The definition of $d_n(z)$ implies that the region Ω,

$\Omega=\{\zeta\in\bar{\mathbb{D}}:|\zeta-z|\leq d_n(z)\}$, contains at least $n+1$ points of spec I. Applying Lemma 1.4 to Ω we obtain

$$|f^{(\nu)}(z)|\underset{\phi,n}{\leq} d_n^{n-\nu}(z)\,\omega(|\phi((1-d_n(z))z)|d_n(z)),\quad \nu=0,\ldots,n. \tag{1.3.1}$$

Now if $a_n(z)\leq \dfrac{4A}{d_n(z)}$ and A is the number defined in Lemma 1.10 then the desired estimate (1.2.11) follows from (1.3.1).

If $a_n(z)>\dfrac{4A}{d_n(z)}$, then (1.2.4) yields (1.2.11).

We now turn to the proof of (1.2.10). We determine A as before and consider two cases

1. $a_n(z)>\dfrac{4A}{d_n(z)}$. Let $\sigma=\dfrac{1}{a_n(z)}$. The region $\tilde{\Omega}=\{\zeta\in\bar{\mathbb{D}}:|\zeta-z|\leq\sigma\}$ cannot contain more than n points from spec I by the definition of $d_n(z)$. Hence there is a k, $1\leq k\leq n+1$ such that the region $\Omega_K=\{\zeta\in\bar{\mathbb{D}}:\dfrac{k\sigma}{n+2}<$ $<|\zeta-z|\leq\dfrac{(k+1)\sigma}{n+2}\}$ does not contain points of spec I. For such a k we put

$$\Omega=\{\zeta\in\bar{\mathbb{D}}:|\zeta-z|\leq\frac{k\sigma}{n+2}+\frac{\sigma}{2(n+2)}\}.$$

Next we put $\mathcal{J}=I|_{\operatorname{spec} I\setminus\Omega}$. Then $I=\mathcal{J}b$, where $b(\zeta)$ is a finite Blaschke product with at most n factors. Let $f_0=f/\mathcal{J}$. Using Lemmas 1.7, 1.10 and the equality

$$f_0^{(\nu)}=\sum_{l=0}^{\nu}C_\nu^l f^{(l)}(1/y)^{(\nu-l)} \tag{1.3.2'}$$

we obtain in our case

$$|f_0^{(\nu)}(\zeta)|\underset{\phi,n}{\leq}\sigma^{n-\nu}\omega(|\phi((1-\sigma)z)|\sigma),\quad \nu=0,\ldots,n. \tag{1.3.2}$$

Now we pick any two points $\zeta_1,\zeta_2\in\Omega$. We have

$$|f_0^{(n)}(\zeta_1)-f_0^{(n)}(\zeta_2)|\leq\sum_{\nu=0}^{n-1}C_n^\nu\,[|f^{(\nu)}(\zeta_1)-f^{(\nu)}(\zeta_2)||(1/y)^{(n-\nu)}(\zeta_1)|+$$

$$+|f^{(\nu)}(\zeta_2)||(1/y)^{(n-\nu)}(\zeta_1)-(1/y)^{(n-\nu)}(\zeta_2)|] + |f^{(n)}(\zeta_1)-f^{(n)}(\zeta_2)|\frac{1}{|y(\zeta_1)|}+$$

$$+|f^{(n)}(\zeta_2)||\frac{1}{y(\zeta_1)}-\frac{1}{y(\zeta_2)}|.$$

Now for $0 \leqslant \nu \leqslant n-1$ with the help of Lemma 1.7 and (1.3.2) we obtain

$$|f^{(\nu)}(\zeta_1)-f^{(\nu)}(\zeta_2)| \leqslant \sigma^{n-\nu-1}\cdot|\zeta_1-\zeta_2|\cdot\omega, \tag{1.3.3'}$$

$$|(1/y)^{(n-\nu)}(\zeta_1)-(1/y)^{(n-\nu)}(\zeta_2)| \leqslant \sigma^{\nu-n-1}\cdot|\zeta_1-\zeta_2|, \tag{1.3.3''}$$

$$|(1/y)^{(n-\nu)}(\zeta_1)| \lesssim \sigma^{\nu-n}, \tag{1.3.3'''}$$

where $\omega = \omega(|\Phi|((1-\sigma)z)|\sigma)$ and hence in (1.3.3) in view Lemma 1.3 we obtain

$$\sum_{\nu=0}^{n-1} \lesssim \frac{|\zeta_1-\zeta_2|\omega}{\sigma} \lesssim (|\Phi(\check{z}_{\zeta_1,\zeta_2})||\zeta_1-\zeta_2|).$$

If now $\nu = n$ then (1.2.11) and (1) imply

$$|f^{(n)}(\zeta_1)-f^{(n)}(\zeta_2)| \leqslant \omega(|\Phi(\check{z}_{\zeta_1,\zeta_2})||\zeta_1-\zeta_2|), \tag{1.3.4}$$

$$|f^{(n)}(\zeta_2)| \leqslant \omega(|\Phi((1-\sigma)z)|\sigma), \tag{1.3.5}$$

and as a consequence of (1.3.3') - (1.3.5) we obtain

$$|f_0^{(n)}(\zeta_1)-f_0^{(n)}(\zeta_2)| \leqslant \omega(|\Phi(\check{z}_{\zeta_1,\zeta_2})||\zeta_1-\zeta_2|), \quad \zeta_1,\zeta_2 \in \Omega, \tag{1.3.6}$$

$$|f_0^{(\nu)}(\zeta)| \lesssim \sigma^{n-\nu}\omega(|\Phi((1-\sigma)z)|\sigma), \quad \nu = 0,\dots,n, \quad \zeta \in \Omega. \tag{1.3.7}$$

We now apply Lemma 1.5 to f_0, Ω and b . This yields

$f_0/b \in \Lambda^n_\omega(\Phi;\Omega)$ (*). But $f_0/b = f/I \stackrel{\text{def}}{=} F$. Applying the Cauchy formula to F at the point $z_0 = (1 - \frac{\sigma}{n+2})z$ we get

$$F^{(\nu)}(z_0) = \frac{\nu!}{2\pi i}\int_{\partial\mathbb{D}} \frac{F(\zeta)}{(\zeta - z_0)^{\nu+1}}d\zeta = \frac{\nu!}{2\pi i}\int_{\partial\Omega} \frac{f_0(\zeta)}{b(\zeta)}\frac{d\zeta}{(\zeta - z)^{\nu+1}}. \tag{1.3.8}$$

The region Ω is choosen so that for $\zeta \in \partial\Omega$ $|b(\zeta)| \geqslant \tilde{C}_n > 0.$

Further $|\zeta - z_0| \geqslant \sigma$, $\zeta \in \partial\Omega$ and therefore (1.3.4) and (1.3.7) imply

$$|F^{(\nu)}(z_0)| \lesssim \sigma^{n-\nu}\omega(|\Phi(z_0)|\sigma), \quad \nu = 0,\dots,n. \tag{1.3.9}$$

Now from (*) and (1.3.9) we deduce

$$|F^{(n)}(z)| \leqslant |F^{(n)}(z_0)| + |F^{(n)}(z) - F^{(n)}(z_0)| \lesssim \omega(|\Phi(z_0)|\sigma), \tag{1.3.10}$$

$$|F^{(\nu)}(z)| \leqslant \sum_{K=0}^{n-\nu} \frac{|F^{(\nu+K)}(z_0)|}{K!}|z - z_0|^K + \frac{1}{(n-K-1)!}\left|\int_{z_0}^{z}(z-t)^{n-\nu-1}(F^{(n)}(t) - F^{(n)}(z_0))dt\right| \leqslant$$

$$\lesssim \sigma^{n-\nu}\omega(|\Phi(z_0)|\sigma), \quad \nu = 0,\dots,n-1. \tag{1.3.11}$$

The inequalities (1.3.10) and (1.3.11) complete the proof in case 1.

CASE 2 . $a_n(z) \leqslant \frac{4A}{d_n(z)}$. We put $\sigma = \frac{1}{4}d_n(z)$ and apply literally the same arguments as in case 1. The only difference is that we use Lemmas 1.5 and 1.7 instead of Lemmas 1.7 and 1.9 after the specification of Ω .

Repeating then the arguments of case 1, we accomplish the proof of the lemma. ●

1.4. Proof of Theorem 1.

Let $f\in\Lambda^n_\omega(\Phi)$, f satisfy (1), I be an inner function, $|I|\leqslant 1$, $F\overset{def}{=}f/I\in C_A$. By theorem A we have $F\in\Lambda^n_\omega(\Phi)$, if only we can check that

$$|F^{(n)}(z_1)-F^{(n)}(z_2)|\lesssim\omega(|\Phi(\check{z}_{z_1,z_2})||z_1-z_2|),\quad z_1,z_2\in\partial\mathbb{D}. \tag{1.4.1}$$

CASE 1. $|z_1-z_2|\geqslant\max\left(\frac{1}{4}d_n(z_1),\frac{1}{4}d_n(z_2)\right)$, $d_n(z)=d_{n,I}(z)$.

Lemma 1.11 implies

$$|F^{(n)}(z_j)|\lesssim\omega(|\Phi((1-d_n(z_j))z_j)|d_n(z_j))\lesssim$$

$$\lesssim\omega(|\Phi(\check{z}_{z_1,z_2})||z_1-z_2|),\qquad j=1,2,$$

$$|F^{(n)}(z_1)-F^{(n)}(z_2)|\lesssim\omega(|\Phi(\check{z}_{z_1,z_2})||z_1-z_2|).$$

CASE 2. $|z_1-z_2|<\max\left(\frac{1}{4}d_n(z_1),\frac{1}{4}d_n(z_2)\right)$.

We can assume $d_n(z_1)\geqslant d_n(z_2)$.

We consider two subcases.

2.1. $|z_1-z_2|\geqslant\max\left(\dfrac{1}{100n\,a_n(z_1)},\dfrac{1}{100n\,a_n(z_2)}\right)$.

Lemma 1.11 implies

$$|F^{(n)}(z_1)|+|F^{(n)}(z_2)|\lesssim\omega\left(\left|\Phi\left(\left(1-\frac{1}{a_n(z_1)}\right)z_1\right)\right|\cdot\frac{1}{a_n(z_1)}\right)+$$

$$+\omega\left(\left|\Phi\left(\left(1-\frac{1}{a_n(z_2)}\right)z_2\right)\right|\cdot\frac{1}{a_n(z_2)}\right)\lesssim\omega(|\Phi(\check{z}_{z_1,z_2})||z_1-z_2|).$$

2.2.

$$|z_1-z_2|<\max\left(\frac{1}{100n\,a_n(z_1)},\frac{1}{100n\,a_n(z_2)}\right).$$

But we have $|z_1-z_2|<\frac{1}{4}d_n(z_1)$, $\quad d_n(z_2)\leqslant d_n(z_1)$.

This implies $d_n(z_2) \geq \frac{3}{4} d_n(z_1)$ since both of the regions $\tilde{\Omega}_1 = \{\zeta \in \bar{\mathbb{D}} : |\zeta - z_1| < d_n(z_1)\}$ and $\tilde{\Omega}_2 = \{\zeta \in \bar{\mathbb{D}} : |\zeta - z_2| < d_n(z_2)\}$ cannot contain more than n points of spec I and since the regions $\tilde{\Omega}_0 = \{\zeta \in \bar{\mathbb{D}} : |\zeta - z_2| < \frac{3}{4} d_n(z_1)\}$ and $(\tilde{\Omega}_1 \setminus \tilde{\Omega}_2) \cup (\tilde{\Omega}_2 \setminus \tilde{\Omega}_1)$ do not intersect, we have in our case

$$\frac{1}{C_{n2}} a_n(z_2) \leq a_n(z_1) \leq C_{n2} a_n(z_2). \qquad (1.4.2)$$

Put

$$\sigma = \min\left(\frac{d_n(z_1)}{4(n+2)}, \frac{1}{100 C_{n2} \cdot n(n+2) a_n(z_1)}\right).$$

Then $(n+2)\sigma \leq \frac{1}{4} d_n(z_1)$. The region $\check{\Omega} = \{\zeta \in \bar{\mathbb{D}} : |\zeta - z_1| \leq |z_1 - z_2| + (n+2)\sigma\}$ cannot contain more then n points of spec I. Hence at least one of the regions

$$\Omega_K = \{\zeta \in \bar{\mathbb{D}} : |z_1 - z_2| + K\sigma < |\zeta - z_1| \leq |z_1 - z_2| + (K+1)\sigma\}, \quad K = 0, \dots, n,$$

does not intersect spec I. For such a K we put

$$\Omega = \{\zeta \in \bar{\mathbb{D}} : |\zeta - z_1| \leq |z_1 - z_2| + (K + \tfrac{1}{2})\sigma\}.$$

The definition of σ and the assumptions of subcase 2.2 imply

$$\frac{\sigma}{2} \leq \operatorname{diam} \Omega \leq C_{n3}\sigma. \qquad (1.4.3)$$

We put

$$y = I\big|_{\operatorname{spec} I \setminus \Omega}, \quad b = \frac{I}{y}.$$

By the construction of Ω the function b is a finite Blaschke product with at most n factors. Let $f_0 = f/y$. Lemma 1.7, (1.4.2) and (1.4.3) imply

$$|(\tfrac{1}{y})^{(\nu)}(\zeta)| \underset{n}{\lesssim} \sigma^{-\nu}, \quad \nu = 0,\dots,n+1, \quad \zeta \in \Omega. \tag{1.4.4}$$

We apply now the Taylor formula and the inequality (1.2.11) to f and obtain

$$|f^{(\nu)}(\zeta)| \underset{n,\Phi}{\lesssim} \sigma^{n-\nu}\omega(|\Phi((1-\sigma)z_1)|\sigma), \quad \nu = 0,\dots,n, \quad \zeta \in \Omega. \tag{1.4.5}$$

Treating (1.4.4) and (1.4.5) similarly to (1.3.3) and (1.2.11) we find

$$|f_0^{(n)}(\zeta_1) - f_0^{(n)}(\zeta_2)| \underset{n,\Phi}{\lesssim} \omega(|\Phi(\check{z}_{\zeta_1,\zeta_2})||\zeta_1-\zeta_2|), \quad \zeta_1,\zeta_2 \in \Omega. \tag{1.4.6}$$

But $f_0|b = f/I = F$. Since f_0 satisfies (1.4.6) and any factor of b satisfies the assumptions of Lemma 1.5 we obtain

$f_0/b \underset{n,\Phi}{\in} \Lambda^n_\omega(\Phi;\Omega)$. Because $z_1, z_2 \in \Omega$ we deduce

$$|F^{(n)}(z_1) - F^{(n)}(z_2)| \leq C_{n\Phi}\,\omega(|\Phi(\check{z}_{z_1,z_2})||z_1-z_2|),$$

Theorem 1 is proved now. ●

§ 2. The multiplication

THEOREM 2. Let $f \in \Lambda^n_\omega(\Phi)$, I be an inner function, $|I| \leq 1$, $f/I \in C_A$ and let the multiplicity of zero $\alpha \in \operatorname{spec} I \cap \mathbb{D}$ of f be at least $n+1$. Then $fI \in \Lambda^n_\omega(\Phi)$.

PROOF. Let $\rho(z) \overset{def}{=\!=} d_{n,I_f}(z)$. The assumption on the zeros of f and the note on ρ imply $dist(z, \operatorname{spec} I_f) \leq \rho(z)$. As in the proof of Theorem 1 it is sufficient to check the following inequality

$$|(fI)^{(n)}(z_1) - (fI)^{n}(z_2)| \lesssim (|\Phi(\check{z}_{z_1,z_2})||z_1-z_2|), \quad z_1,z_2 \in \partial\mathbb{D}. \tag{2.0}$$

We can now apply Lemma 1.7 with $d(z) = \operatorname{dist}(z, \operatorname{spec} I)$ to estimate I. To estimate f we can use (1.2.11) where μ_ν is defined with the help of $d_{n,I_f}(z) = \rho(z) \leq d(z)$, and $a(z) = |I'(z)|$, where in view of the assumption $|I_f/I| \leq 1$ and the assumption on multiplicity of zeros of f we obtain $a(z) \leq a_{n,I_f}(z)$. We write

$$|fI|^{(n)} = \sum_{\nu=0}^{n} C_n^\nu f^{(\nu)} I^{(n-\nu)}$$

and consider two main cases as in the proof of Theorem 1.

CASE 1. $|z_1 - z_2| \geq \max\left(\frac{1}{4} d(z_1), \frac{1}{4} d(z_2)\right)$.

We use Lemmas 1.7, 1.11 and Leibnitz formula to estimate $|(fI^{(n)}(z_1)| + |(fI)^{(n)}(z_2)|$ it has been made in subsection 1.4.

CASE 2. $|z_1 - z_2| < \max\left(\frac{1}{4} d(z_1), \frac{1}{4} d(z_2)\right)$.

SUBCASE 2.1. $|z_1 - z_2| \geq \max\left(\frac{1}{100 n\, a(z_1)}, \frac{1}{100 n\, a(z_2)}\right)$.

We deal as in subcase 2.1 of the proof of Theorem 1.

SUBCASE 2.2. $|z_1 - z_2| < \max\left(\frac{1}{100 n\, a(z_1)}, \frac{1}{100 n\, a(z_2)}\right)$.

We put $\Omega = \check{\Omega}$ in the notation of 1.4. Then for

$$\sigma = \min\left(\frac{d(z_1)}{4(n+2)}, \frac{1}{100 n (n+2) C_{n2}\, a(z_1)}\right),$$

where C_{n2} is taken from (1.4.2), we obtain the inequalities

$$|I^{(\nu)}(\zeta)| \underset{n}{\lesssim} \sigma^{-\nu}, \quad \nu = 0, \ldots, n+1, \quad \zeta \in \Omega,$$

$$|f^{(\nu)}(\zeta)| \underset{n,\phi}{<} \sigma^{n-\nu} \omega(|\phi((1-\sigma)z_1)|\sigma), \quad \nu = 0, \ldots, n, \ \zeta \in \Omega,$$

Using these inequalities and the Leibnitz formula we obtain (2.0). ●

THEOREM 3. Let ω be an arbitrary modulus of continuity. There exist a function $f \in A^\infty$ and a Blaschke product B such that $f/B \in A^\infty$, but $fB \notin \Lambda^1_\omega$ and $fI_f \notin \Lambda^1_\omega$.

PROOF. If $f/B \in A^\infty$, then $\mathcal{J} \overset{def}{=} I_f/B$ is a bounded inner function. If we would have $fI_f = fB\mathcal{J} \in \Lambda^1_\omega$, then Theorem 1 would imply

$fB \in \Lambda^1_\omega$.

It means that theorem will be proved if we find a function f and a Blaschke product B with properties required.

LEMMA 2.1. Let $a \in \mathbb{D}$, $a \neq 0$, $h(z) = \frac{(z-a)^2}{1-\bar{a}z}$. Then

$$|h'(\frac{a}{|a|} - h'(a))| = |h'(\frac{a}{|a|})| = 2+|a|,$$

$$|h(\frac{a}{|a|})| = 1-|a|. \quad \bullet \tag{2.1}$$

We define now the module of the following outer function g_ν, $\nu = 2,3,\dots$

$$\log|g_\nu(e^{i\theta})| = \begin{cases} -\left(\frac{1}{\sqrt{|\theta|}} + \frac{1}{\sqrt{\pi+\theta}}\right), & -\pi < \theta < 0; \\ -\left(\frac{1}{\sqrt{\theta - \frac{\pi}{2^{m+1}}}} + \frac{1}{\sqrt{\frac{\pi}{2^m} - \theta}}\right), & \theta \in \left(\frac{\pi}{2^{m+1}}, \frac{\pi}{2^m}\right), m \leq \nu-2 \text{ or } m \geq \nu+1; \\ -\left(\frac{1}{\sqrt{\theta - \frac{\pi}{2^{\nu+1}}}} + \frac{1}{\sqrt{\frac{\pi}{2^{\nu-1}} - \theta}}\right), & \theta \in \left(\frac{\pi}{2^{\nu+1}}, \frac{\pi}{2^{\nu-1}}\right). \end{cases}$$

Then as in [2] it is easy to check $g_\nu \in A^\infty$ and

$$\|g_\nu\|_{\Lambda^n} \leq C_n, \quad \nu = 2,3,\dots, \quad n = 1,2,\dots, \tag{2.2}$$

where C_n do not depend on ν and

$$|g_\nu^{(j)}(z)| \leq \tilde{C}_j \operatorname{dist}^{2(j+2)}(z, E_\nu), \; j = 1,2,\dots, \; \nu = 2,3,\dots, \; E_\nu = \bigcup_{\substack{m \geq 0 \\ m \neq \nu}} \{e^{\pi i 2^{-m}}\}, \tag{2.3}$$

where $\tilde{C}_j$ do not depend on ν . Let $\alpha_\nu = (1-\varepsilon_\nu) e^{\frac{\pi i}{2^\nu}}$, where $\varepsilon_\nu \downarrow 0$.

We put

$$B_\nu(z) = \prod_{\substack{k=2 \\ k \neq \nu}} \frac{\bar{\alpha}_k}{|\alpha_k|} \cdot \frac{\alpha_k - z}{1 - \bar{\alpha}_k(z)}, \quad B(z) = \prod_{k=2}^{\infty} \frac{\bar{\alpha}_k}{|\alpha_k|} \cdot \frac{\alpha_k - z}{1 - \bar{\alpha}_k z}$$

and

$$f(z) = \sum_{\nu=2}^{\infty} \frac{1}{\nu^2} g_\nu(z) B_\nu(z) (z - \alpha_\nu). \tag{2.4}$$

Clearly, $f(\alpha_\nu)=0,\ \nu\geq 2$. Then (2.2), (2.3) and $|B_\nu^{(j)}(z)|\leq C_j^0\, dist^{-2j}(z, \bigcup_{\substack{m\geq 0\\ m\neq\nu}}\{e^{-\pi i 2^{-m}}\}$ [15] imply

$$\|g_\nu B_\nu\|_{\Lambda^n}\leq C_{n0},\quad \nu\geq 2,\quad n\geq 1, \tag{2.5}$$

and then

$$\|f\|_{\Lambda^n}\leq\sum_{\nu\geq 2}\frac{1}{\nu^2}\|g_\nu B_\nu\|_{\Lambda^n}\cdot\|z-\alpha_\nu\|_{\Lambda^n}<\infty,\quad n\geq 1,$$

This means that $f\in A^\infty$ and therefore $f/B\in A^\infty$ by THEOREM 1. We now estimate $v_{\nu_0}=(fB)'(\alpha_{\nu_0})-(fB)'(\alpha^0_{\nu_0})$. We have the inequality

$$|(fB)'(\alpha_{\nu_0})-(fB)'(\alpha^0_{\nu_0})|=\Big|\sum_{\nu\geq 2}\frac{1}{\nu^2}\Big[(g_\nu(z)B_\nu(z)B(z)(z-\alpha_\nu))'_{z=\alpha^0_{\nu_0}}-$$

$$-(g_\nu(z)B_\nu(z)B(z)(z-\alpha_\nu))'_{z=\alpha^0_{\nu_0}}\Big]\Big|=\frac{1}{\nu_0^2}\Big|\Big(g_{\nu_0}(z)B^2_{\nu_0}(z)\frac{(z-\alpha_{\nu_0})^2}{1-\bar\alpha_{\nu_0}z}\Big)_{z=\alpha^0_{\nu_0}}\Big|=$$

$$=\frac{1}{\nu_0^2}\Big|(g_{\nu_0}B^2_{\nu_0})'(\alpha^0_{\nu_0})\frac{(\alpha^0_{\nu_0}-\alpha_{\nu_0})^2}{1-\bar\alpha_{\nu_0}\alpha^0_{\nu_0}}+g_{\nu_0}(\alpha^0_{\nu_0})B^2_{\nu_0}(\alpha^0_{\nu_0})\Big(\frac{(z-\alpha_{\nu_0})^2}{1-\bar\alpha_{\nu_0}z}\Big)'_{z=\alpha^0_{\nu_0}}\Big|\geq$$

$$\geq\frac{1}{\nu_0^2}\left(2|g_{\nu_0}(\alpha_\nu)|-C_0\varepsilon_{\nu_0}\right)$$

and C_0 does not depend on ν_0 . We choose $\{\varepsilon_\nu\}_{\nu\geq 2}$ so that

$$C_0\varepsilon_\nu\leq|g_\nu(\alpha^0_\nu)|,\quad \nu\geq 2, \tag{2.7}$$

$$L_\nu\overset{def}{=\!=}\frac{|g_\nu(\alpha^0_\nu)|}{\nu^2\omega(\varepsilon_\nu)}\xrightarrow[\nu\to\infty]{}\infty\,. \tag{2.8}$$

Then (2.7)- (2.8) imply

$$|v_{\nu_0}|\geq\frac{1}{\nu_0^2}|g_{\nu_0}(\alpha^0_{\nu_0})|=L_{\nu_0}\omega(\varepsilon_{\nu_0}),$$

and hence $Bf\notin\Lambda^1_\omega$ because otherwise we would have $|v_{\nu_0}|\leq L\omega(\varepsilon_{\nu_0})$, $L<\infty$. Theorem 3 is proved now. ●

§ 3. The absence of the (F)-property

Let $a>0$, $c>0$ and $\sigma>0$. We put

$$1\leq b_n\leq c(n+1)^{\sigma},\ n\geq 0,\quad {}_{\sigma}b \overset{def}{=\!=} \{b_n\}_{n\geq 0};$$

$$\ell^p_A({}_{\sigma}b) \overset{def}{=\!=} \{f(z)=\sum_{n\geq 0}\hat{f}(n)z^n : \big(\sum_{n\geq 0} b_n|\hat{f}(n)|^p\big)^{1/p}<\infty,\quad f\in H^1\}.$$

THEOREM 4. For $1\leq p\leq\infty, p\neq 2$ the class $\ell^p_A({}_{\sigma}b)$ does not possess the (F)-property.

Proof. We consider only the case $p>2$, since the case $1\leq p<2$ is simpler. Let

$$S_{\delta}(z)\overset{def}{=\!=}\exp\Big(\delta\frac{z+1}{z-1}\Big),\quad I_{\delta}\overset{def}{=\!=}\{f\in H^1 : f/S_{\delta}\in H^1\}.$$

We find a function $f\in \ell^p_A({}_{\sigma}b)\cap I_{\delta_0}$, $\delta_0=\frac{\pi^2}{32}$, such that $f/S_{\delta_0}\notin \ell^p_A({}_{\sigma}b)$

To this end due to S.Banach's theorem it is sufficient to constract a sequence of functions $f_n\in \ell^p_A({}_{\sigma}b)\cap I_{\delta_0}$ with the properties $\|f_n\|_{\ell^p_A({}_{\sigma}b)}\leq C$, $\|f_n/S_{\delta_0}\|_{\ell^p_A({}_{\sigma}b)}\to\infty$.

LEMMA 3.1. [31] Ch.5. Let $0<\mu<\frac{1}{4}$, $\alpha>0$, $N\geq 1$

$$g_N(\theta)=\sum_{n=1}^{N}\frac{\cos(\alpha\sqrt{n}+\frac{\pi}{4})}{n^{1/4+\mu}}e^{in\theta}. \tag{3.1}$$

Then

$$\int_{-\pi}^{\pi}|g_N(\theta)|\,d\theta\leq C,$$

where C does not depend on N.

LEMMA 3.2. Let $a>1$, $M\geq a^4$, $0<\mu<\frac{1}{4}$, $\alpha>0$. Then for some $C_1>0$ not depending on a and M for $a>a_0(\alpha,\mu)$, we have

$$\Big|\sum_{n=1}^{M}\frac{\cos(\alpha\sqrt{n}-\frac{\pi}{4})}{n^{3/4}}\cdot\frac{\cos(\alpha\sqrt{n+a}-\frac{\pi}{4})}{(n+a)^{1/4+\mu}}\Big|\geq\frac{C_1}{a^{2\mu}}. \tag{3.1'}$$

PROOF. The mean-value theorem yields

$$\left|\sum_{n=1}^{M}\frac{\cos(\alpha\sqrt{n}+\frac{\pi}{4})}{n^{3/4}}\cdot\frac{\cos(\alpha\sqrt{n+a}+\frac{\pi}{4})}{(n+a)^{1/4+\mu}}-\int_0^M\frac{\cos(\alpha\sqrt{x}+\frac{\pi}{4})\cos(\alpha\sqrt{x+a}+\frac{\pi}{4})}{x^{3/4}\,(x+a)^{1/4+\mu}}dx\right|\leq\frac{c_2}{a^{1/4+\mu}},$$

where c_2 does not depend on a and M. Further,

$$\int_0^M=\frac{1}{a^{\mu}}\int_0^{M/a}\frac{\cos(\alpha\sqrt{a}\sqrt{t}+\frac{\pi}{4})\cos(\alpha\sqrt{a}\sqrt{t+1}+\frac{\pi}{4})}{t^{3/4}(t+1)^{1/4+\mu}}dt=$$

$$=\frac{1}{2a^{\mu}}\int_0^{M/a}\frac{\cos(\alpha\sqrt{a}(\sqrt{t}-\sqrt{t+1}))}{t^{3/4}(t+1)^{1/4+\mu}}dt-$$

$$-\frac{1}{2a^{\mu}}\int_0^{M/a}\frac{\sin\alpha\sqrt{a}(\sqrt{t}+\sqrt{t+1})}{t^{3/4}(t+1)^{1/4+\mu}}dt=I_1-I_2.$$

We make change of variables in the integral I_1. Then

$$t=\left[\frac{1}{2}\left(v-\frac{1}{v}\right)\right]^2,$$

$$\frac{dt}{t^{3/4}(t+1)^{1/4+\mu}}=-2^{1+2\mu}\frac{dv}{v^{1-2\mu}}\frac{(1+v^2)^{1/2-2\mu}}{(1-v)^{1/2}}=-2^{1+2\mu}\frac{dv}{v^{1-2\mu}}+g(v)\,dv,$$

where $g\in C^1[0,\frac{1}{2}]$, $g(0)=g'(0)=0$. Hence

$$I_1=\frac{c}{a^{\mu}}\int_{\frac{1}{\sqrt{\frac{M}{a}}-\sqrt{\frac{M}{a}+1}}}^{1}\frac{\cos\sqrt{a}\,v}{v^{1-2\mu}}dv+\frac{1}{2a^{\mu}}\int_{\frac{1}{\sqrt{\frac{M}{a}}-\sqrt{\frac{M}{a}+1}}}^{1/2}\cos\alpha\sqrt{a}\,v\cdot g(v)\,dv+$$

$$+\frac{2^{2\mu}}{a^{\mu}}\int_{1/2}^{1}\frac{(1+v^2)^{1/2-2\mu}\cos\alpha\sqrt{a}\,v}{v^{1-2\mu}(1-v^2)^{1/2}}dv=J_1+J_2+J_3. \tag{3.2}$$

Integrating J_2 by parts we find $|J_2|\leq c_3a^{-\mu-\frac{1}{2}}$. We make a change of variables $\sqrt{a}\,v=u$ in the integral J_1. Then we obtain

$$\mathcal{J}_1=\frac{2^{2\mu}}{a^{2\mu}}\int\limits_{\frac{\sqrt{a}}{\sqrt{\frac{M}{a}}+\sqrt{\frac{M}{a}+1}}}^{\sqrt{a}/2}\frac{\cos\alpha\sqrt{u}}{u^{1-2\mu}}\,du=\frac{2^{2\mu}}{a^{2\mu}}\mathcal{J}_4. \tag{3.3}$$

For $a\geq a_0(\mu,\alpha)$ and $M\geq a^4$, $|\mathcal{J}_4|\geq c_4>0$, i.e. $|\mathcal{J}_1|\geq c_5a^{-2\mu}$. The integrals $\mathcal{J}_3$ and I_2 can be theated like (3.2) and (3.3), $|\mathcal{J}_3|,|I_2|\leq c_6a^{-\mu-1/4}$, which together with (3.2) proves the lemma.

LEMMA 3.3. Let $1\leq b_n\leq Cn^a$, $n\geq 1$, $a>0$,

$$\alpha_\nu\geq 0,\quad A\geq 1,\quad \beta_\mu,\quad 0\leq\mu\leq B,\quad B\geq 1,$$

$$\alpha=\sum_{\nu=0}^{A}\alpha_\nu>0,\quad \beta=\sum_{\mu=0}^{B}\beta_\mu,\quad 2L=\frac{\beta}{\alpha},\quad L\geq 1.$$

Then

$$\varlimsup_{n\to\infty}\frac{\sum_{\mu=0}^{B}b_{n+\mu}\beta_\mu}{\sum_{\mu=0}^{A}b_{n+\mu}\alpha_\mu}\geq 0,8\,L. \tag{3.4}$$

PROOF. Suppose (3.4) is false then for $n\geq n_0$ we have

$$\sum_{\mu=0}^{B}b_{n+\mu}\beta_\mu\leq 0,9\,L\sum_{\nu=0}^{A}b_{n+\nu}\alpha_\nu. \tag{3.5}$$

Summing (3.5) from $n=N-A-B$ to $n=N+Q+A+B$. we find

$$\sum_{n=N-A-B}^{N+Q+A+B}\sum_{0}^{B}b_{n+\mu}\beta_\mu\leq 0,9\sum_{n=N-A-B}^{N+Q+A+B}\sum_{\nu=0}^{A}b_{n+\nu}\alpha_\nu$$

$$\sum_{n=N-A-B}^{N-1}b_n\sum_{\substack{\mu\leq B\\ \mu\leq n-N+A+B}}\beta_\mu+\beta\sum_{n=N}^{N+Q-1}b_n+\sum_{n=N+Q}^{N+Q+A+B}b_n\sum_{\substack{\mu\leq B\\ \mu\leq N+Q+A+B-n-1}}\beta_\mu\leq$$

$$\leq 0,9L\sum_{n=N-A-B}^{N-1}b_n\sum_{\substack{\nu\leq A\\ \nu\leq n-N+A+B}}\alpha_\nu+0,9L\alpha\sum_{n=N}^{N+Q-1}b_n+0,9L\sum_{n=N+Q}^{N+Q+A+B}b_n\sum_{\substack{\nu\leq A\\ \nu\leq N+Q+A+B-n-1}}\alpha_\nu. \tag{3.6}$$

Since $\alpha_\nu, \beta_\nu \geq 0$, we use that (3.6) implies

$$2L\alpha\sum_{N}^{N+Q-1} b_n \leq 0{,}9L\alpha\sum_{N}^{N+Q-1} b_n + 0{,}9L\alpha\sum_{N-A-B}^{N-1} b_n + 0{,}9L\alpha\sum_{N+Q}^{N+Q+A+B} b_n, \tag{3.7}$$

or

$$\sum_{N}^{N+Q-1} b_n \leq \sum_{N-A-B}^{N-1} b_n + \sum_{N+Q}^{N+Q+A+B} b_n$$

for any $N \geq n_0,\ Q \geq 1$. Let $T = \sum_{n_0-A-B}^{n_0-1}$. Then since $b_n \geq 1$ we have

$$\sum_{n_0}^{n_0+Q-1} b_n \leq (T+1)\sum_{n_0+Q}^{n_0+Q+A+B} b_n . \tag{3.8}$$

Putting in (3.8) $Q = A+B,\ 2(A+B), \ldots, \nu(A+B)$ and $\Sigma_\nu = \sum^{n_0+\nu(A+B)-1} b_n$ we find from (3.8)

$$\Sigma_\nu \leq (T+1)(\Sigma_{\nu+1} - \Sigma_\nu),$$

$$\Sigma_{\nu+1} \geq \left(\frac{T+2}{T+1}\right)^\nu \Sigma \geq \left(\frac{T+2}{T+1}\right)^\nu, \quad \nu \geq 1,$$

$$\Sigma_{\nu+1} - \Sigma_\nu \geq \frac{1}{T+1}\cdot\left(\frac{T+2}{T+1}\right)^{\nu-1}, \quad \nu \geq 2. \tag{3.9}$$

But the assumption gives

$$\Sigma_{\nu+1} - \Sigma_\nu = \sum_{n_0+\nu(A+B)}^{n_0+(\nu+1)(A+B)-1} b_n \leq c(A+B)[n_0+\nu(A+B)]^a. \tag{3.10}$$

The inequalities (3.9) and (3.10) being contradictory, it follows that our assumption was wrong, i.e. (3.4) holds. ●

LEMMA 3.4. Let $\{\gamma_n^0\}_n$ be a finite sequence,

$$v(\zeta)=\sum_n \frac{\gamma_n^0}{\zeta-n}.$$

Let

$$|v(\zeta)|=O(|\zeta|^{-4}), \quad \zeta\to\infty, \tag{3.11}$$

and let

a) $\gamma^0_n=0$ for $n=4N^2$, $N=0,1,\dots$

or

b) $\gamma^0_n=0$ for $n=(2N-1)^2$

We put $\frac{\sin\frac{\pi\sqrt{z}}{2}}{\sqrt{z}}=f_a(z)$, $\cos\frac{\pi\sqrt{z}}{2}=f_b(z)$ and put in the case a)

$$\hat{\gamma}(n)=\begin{cases}\gamma^0_n/f_a(n), & n\neq 4N^2;\\ v(4N^2)/f_a'(4N^2), & n=4N^2,\ N=0,1,\dots;\end{cases}$$

while in the case b)

$$\hat{\gamma}(n)=\begin{cases}\gamma^0_n/f_b(n), & n\neq (2N-1)^2;\\ v((2N-1)^2)/f_B'((2N-1)^2), & n=(2N-1)^2,\ N=1,2,\dots,\end{cases}$$

Let finally

$$\gamma(z)=\sum_{n\geq 0}\hat{\gamma}(n)z^n, \quad z\in\mathbb{D}.$$

Then

$$\gamma\in I_{\delta_0}, \quad \delta_0=\frac{\pi^2}{32}.$$

PROOF [12], [13] . ●

PROOF OF THE THEOREM. We fix p, $2<p<\infty$ (the case $p=\infty$ is simpler) and take μ, $0<\mu<1/4$ such that $2p\mu<1$, $p\mu+\frac{p}{4}>1$. Now we fix a number $a>1$ to which we could apply Lemma 3.2. We put $2L\overset{def}{=}\sum_1^a n^{-2p\mu}/\sum_1^\infty n^{-p(\mu+1/4)}>2$ and $m=2a^4$, $M=a^8$, $y=[16(\sigma+1)]$.

We introduce the sequences $\{\alpha_\nu\}$, $\{\beta_\nu\}$, $\{\gamma_\nu\}$ as follows:

$$\alpha_\nu^0 = \frac{\cos\left(\frac{\pi}{2}\sqrt{\nu}-\frac{\pi}{4}\right)}{\nu^{1/2+\mu}}, \qquad 1 \leqslant \nu \leqslant m,$$

$$\gamma_\nu = (-1)^K C_Y^K \alpha_\mu^0, \quad \nu = MK+\mu, \quad 1 \leqslant \mu \leqslant m, \quad K = 0,1,\ldots,Y,$$

$$\gamma_\nu = 0, \quad MK+m < \nu \leqslant M(K+1), \quad K = 0,\ldots,Y-1, \tag{3.11'}$$

where $C_Y^K = \binom{Y}{K}$ are binomial coefficients

$$\alpha_\nu = \nu^{-p(\frac{1}{4}+\mu)}, \quad 1 \leqslant \nu \leqslant m,$$

$$\alpha_\nu = (C_Y^K)\alpha_\mu, \quad \nu = MK+\mu, \quad 1 \leqslant \mu \leqslant m, \quad K = 0,1,\ldots,Y;$$

$$\alpha_\nu = 0, \quad MK+m < \nu \leqslant M(K+1), \quad K = 0,\ldots,Y-1;$$

$$\beta_\nu = \nu^{-2p\mu}, \quad 1 \leqslant \nu \leqslant a.$$

To the sequences $\{\alpha_\nu\}$ and $\{\beta_\nu\}$ we apply Lemma 3.3 and find the sequence $\{n_K\}$, $n_K \longrightarrow \infty$ such that

$$\sum_{\nu=1}^{a} b_{n_K+\nu}\,\beta_\nu \geqslant \frac{L}{2}\sum_{\mu=1}^{MY+m} b_{n_K+\mu}\,\alpha_\mu . \tag{3.12'}$$

We split the set of natural numbers into t wo subsets.

$$\mathbb{A} \overset{def}{=} \{n : |n-4N^2| \leqslant 2N\}, \quad \mathbb{B} = \mathbb{N}\setminus\mathbb{A}.$$

One of these sets, say $\mathbb{I}$, contains infinitely many of $\{n_K\}$. We choose n_K so that $n_K > (2aY)^{32}$ and denote $n = n_K$ for brevity. The zeros of f_a lie in $\mathbb{B}$ while the zeros of f_b lie in $\mathbb{A}$. Let f denote that function of f_a and f_b whose zeros lie in $\mathbb{N}\setminus\mathbb{I}$. We put

$$\gamma_{n+\mu}^0 = \gamma_\mu f(n), \quad 1 \leqslant \mu \leqslant MY+m, \quad \gamma_\eta^0 = 0, \quad \eta \leqslant n, \quad \eta > n+MY+m,$$

and construct the functions $v(\zeta)$ and $\gamma(z)$ as in Lemma 3.4. The latter is possible since

$$|v(\zeta)| = O(|\zeta|^{-y}), \quad \zeta \to \infty. \tag{3.12}$$

Lemma 3.4 implies $\gamma \in I_{\delta_0}$ (*).

For any zero ξ, $\xi \geqslant 1$ of f we have

$$|f'(\xi)| \succ \xi^{-1}, \tag{3.13}$$

and the choice of n and f for $1 \leqslant \mu \leqslant B$, $B = My + m$ yields

$$\left|\frac{f(n)}{f(n+\mu)} - 1\right| = \left|\frac{O(f'(t)\mu)}{f(n+\mu)}\right| \leqslant \frac{\mu}{\sqrt{n}} \leqslant \frac{a^8}{a^{16}} = a^{-8}. \tag{3.14}$$

The conditions (3.12) - (3.14) yield

$$\|\gamma\|_{H^1} \leqslant c < \infty, \tag{3.15$_1$}$$

$$\sum_{m \notin [n+1, n+B]} b_m |\hat{\gamma}_m|^p \leqslant C \sum_{m \geqslant 1} m^{\sigma} m^{(1-y)p} \leqslant C_1, \tag{3.15$_2$}$$

$$\sum_{m=n+1}^{n+B} b_m |\hat{\gamma}_m|^p \leqslant \sum_{\mu=1}^{B} b_{n+\mu} d_\mu. \tag{3.15$_3$}$$

For the function $S_{\delta_0}(z) = \exp\left(\delta_0 \frac{z+1}{z-1}\right)$ we have the following expansion [12]

$$\overline{S_{\delta_0}(e^{i\theta})} = C_{\delta_0} \sum_{\nu=0}^{\infty} \left[\frac{\cos\left(\frac{\pi}{2}\sqrt{\nu} + \frac{\pi}{4}\right)}{(\nu+1)^{3/4}} + O\left((\nu+1)^{-5/4}\right)\right] e^{-i\nu\theta}. \tag{3.16}$$

Since

$$\sum_{\nu=1}^{\infty} \frac{1}{\nu^{5/4}} \cdot \frac{1}{(\nu+a)^{1/4+\mu}} = O\left(a^{-\frac{1}{4}-\mu}\right), \quad a \geqslant 1, \tag{3.17}$$

(3.1), (3.11), (3.16) and (3.17) imply

$$|\gamma \hat{\bar{S}}(m)| \geqslant c\nu^{-2\mu}, \quad m = n+\nu, \quad 1 \leqslant \nu \leqslant a, \tag{3.18}$$

and hence from (3.12') and (3.18) we obtain

$$\sum_{m\geqslant 0} b_m |\gamma \hat{\bar{S}}(m)|^p \geqslant c_1 \sum_{\nu=1}^{a} b_{n+\nu} \nu^{-2p\mu} = c_1 \sum_{\nu=1}^{a} b_{n+\nu} \beta_\nu \geqslant$$

$$\geqslant \frac{c_1 L}{2} \sum_{\nu=1}^{B} b_{n+\nu} d_\nu \geqslant c_1' L. \tag{3.19}$$

Finally with the help of (3.15), (*) and (3.19) we conclude

$$\gamma \in I_{\delta_0} \cap \ell_A^p({}_\sigma b).$$

We have

$$\|\gamma / S_{\delta_0}\|_{\ell_A^p({}_\sigma b)} = \|\gamma \bar{S}_{\delta_0}\|_{\ell_A^p({}_\sigma b)} \geqslant cL \|\gamma\|_{\ell_A^p({}_\sigma b)},$$

i.e. $\ell_A^p({}_\sigma b)$ does not possess the (F)-property if $p>2$. In the case $1<p<2$ we do not need Lemma 3.1 since we have $b_n \geqslant 1$ $\ell_A^p({}_\sigma b) \subset \ell_A^2({}_\sigma b) \subset H^2$, and instead of Lemma 3.2 we use the following statement.

LEMMA 3.2. Let $\frac{1}{2} \leqslant \mu < 1$. Then

$$\left|\sum_{\nu=1}^{a} \frac{\cos(d\sqrt{\nu} + \frac{\pi}{4})}{\nu^{3/4}} \cdot \frac{1}{(a-\nu)^\mu}\right| \succ a^{-\frac{1}{4} - \frac{\mu}{2}}, \quad a \geqslant 2.$$

The case $p=1$ we treat as in [50]. The theorem is proved now. ●

CHAPTER II

MODULI OF ANALYTIC FUNCTIONS SMOOTH UP TO THE BOUNDARY

§ 1. The class Λ^α.

In this chapter we use the following notation:

for any arc $I \subset \partial\mathbb{D}$ (for a segment $I \subset \mathbb{R}$) and for a function $f \in C(I)$ we put

$$M_f(I) \overset{def}{=} \max_I |f|;$$

if $z \in \mathbb{D}$, $z \neq 0$ then $z_0 \overset{def}{=} z/|z|$, $\rho = 1-|z|$,

$$\gamma(z) \overset{def}{=} \{\zeta \in \partial\mathbb{D} : |\arg \zeta/z| \leqslant 2\arcsin \rho\}, \quad M_f(z) \overset{def}{=} M_f(\gamma(z));$$

if $\zeta = x+iy \in \Pi$ then $\gamma(\zeta) \overset{def}{=} [x-y, x+y]$;

for $f \in C(\mathbb{R})$, $\zeta \in \Pi$ $M_f(\zeta) \overset{def}{=} M_f(\gamma(\zeta))$.

THEOREM 5. a) Let $f \in \Lambda^\alpha$, α is not an integer. If for a point $z \in \mathbb{D}$ $M_f(z) \geqslant \rho^\alpha$ then we have

$$I_M(f;z) \overset{def}{=} \int_{\partial\mathbb{D}} |\log| \frac{M_f(z)}{f(\zeta)} || \frac{1-|z|^2}{|\zeta - z|^2} |d\zeta| \leqslant B_f, \tag{1}$$

where B_f does not depend on z ;

b) Let $F \in C^\alpha(\partial\mathbb{D})$ and let for any point $z \in \mathbb{D}$ for which $M_f(z) \geqslant$ $\geqslant \rho^\alpha$, $I_M(F;z) \leqslant B_F$, B_F does not depend on z . Then $e^F \in \Lambda^\alpha$.

We mention a corollary of Theorem 5 (see § 4 below), which is a strong form of the Havin-Shamoyan-Carleson-Yacobs theorem for arbitrary α . Namely let $F \in C^\alpha(\partial\mathbb{D}), \alpha \neq 2n$, F satisfies the condition $\int_{\partial\mathbb{D}} \log|F||d\zeta| > -\infty$.

Then $_eF \in \Lambda^{d/2}$.

1.0. Technical preparations.

LEMMA 1.1. Let $\mathcal{D} \subset \mathbb{C}$ be a continuum, $d = \mathrm{diam}\, \mathcal{D}$, $K = \mathbb{D}(w_0, c_1 d)$, $\mathrm{dist}(K, \mathcal{D}) < c_2 d$. Let P be a polynomial, $\deg P = n$, which has $n-m$ zeros $\zeta_1, \ldots$ $\ldots, \zeta_{n-m}$ outside K, a be his senior coefficient,

$$Q = |a| \prod_{\nu=1}^{n-m} |w_0 - \zeta_\nu|.$$

Then

$$\max_{\mathcal{D}} |P| \underset{c_1, c_2, n}{\asymp} Q d^m.$$

PROOF is easy. ●

LEMMA 1.2. Let P be a polynomial, $\deg P = N$, $z \in \mathbb{D}$, $z_0 = z/|z|$, $v(\zeta) = \max_{\mathbb{D}(z_0, |\zeta - z_0|)} |P(t)|$, $M = \max_{\substack{\zeta \in \partial\mathbb{D} \\ \zeta \in \mathbb{D}(z_0, \rho)}} |P(\zeta)|$ and let $n, \tau, a > 0$ be numbers. Then

$$\exp \frac{1}{2\pi} \int_{\partial\mathbb{D}} \log \left[v(\zeta) + a(|\zeta - z_0| + \rho)^{n+\tau} \rho^{-\tau} \right] \cdot \frac{1-|z|^2}{|\zeta - z|^2} |d\zeta| \underset{N,n,\tau}{\lesssim} M + a\rho^n. \qquad (1.0.1)$$

PROOF. Let $M_0 = v(\rho)$. Lemma 1.1 yields $M \underset{N}{\asymp} M_0$.

By Schwarz lemma we have

$$v(\zeta) \leq M_0 \left(\frac{|\zeta - z_0| + \rho}{\rho} \right)^N,$$

hence

$$v(\zeta) + \frac{(|\zeta - z_0| + \rho)^{n+\tau}}{\rho^\tau} \lesssim \left(\frac{|\zeta - z_0| + \rho}{\rho} \right)^{N+\tau+n} (M + a\rho^n)$$

and Lemma 5 [53] gives

$$(1.0.1) \underset{N}{\lesssim} \left[\exp \frac{N+\tau+n}{2\pi} \int_{\partial\mathbb{D}} \log \frac{|\zeta - z_0| + \rho}{\rho} \cdot \frac{1-|z|^2}{|\zeta - z|^2} |d\zeta| \right] (M + a\rho^n) \underset{N,\tau,n}{\lesssim} M + a\rho^n. \quad ●$$

LEMMA 1.3. Let P be a polynomial, $\deg P = n$. Then

$$\int_{\partial\mathbb{D}} \left| \log \left| \frac{P(\zeta)}{M_P(z)} \right| \right| \frac{1-|z|^2}{|\zeta - z|^2} |d\zeta| \underset{n}{\lesssim} 1.$$

PROOF. Let $P(\zeta)=a\prod_{\nu=1}(\zeta-\zeta_\nu)$. Lemma 1.1 yields

$$M_p(z)\underset{n}{\asymp}|a|\prod_{\nu=1}(|z_0-c_\nu|+\rho).$$

Hence

$$\int_{\partial\mathbb{D}}\left|\log\left|\frac{P(\zeta)}{M_p(z)}\right|\right|\frac{1-|z|^2}{|\zeta-z|^2}|d\zeta|\leqslant\sum_{\nu=1}^{n}\int_{\partial\mathbb{D}}\left|\log\left|\frac{\zeta-\zeta_\nu}{|\zeta_\nu-z_0|+\rho}\right|\right|\frac{1-|z|^2}{|\zeta-z|^2}|d\zeta|+C_n\leqslant C_{n1}$$

since, as can easily be seen, we have a uniform estimate

$$\int_{\partial\mathbb{D}}\left|\log\left|\frac{\zeta-\zeta_0}{|z_0-\zeta_0|+\rho}\right|\right|\frac{1-|z|^2}{|\zeta-z|^2}|d\zeta|\leqslant const,\quad \zeta_0\in\mathbb{C}.\quad\bullet$$

1.1. Proof of part a).

LEMMA 1.4. Let P be a polynomial, $\deg P=n$, $Q={}_eP$.

Then

$$|Q(z)|\asymp M_p(z).$$

It is an easy consequence of Lemma 1.1. ●

LEMMA 1.5. Let $f\in\Lambda^\alpha$, $n<\alpha<n+1$, $|f^{(n)}(\zeta)-f^{(n)}(\xi)|\leqslant\leqslant|\zeta-\xi|^{\alpha-n}$, $\zeta,\xi\in\bar{\mathbb{D}}$, P be the Taylor polynomial of f at z_0,

$$B=I_\rho(f;z)\overset{def}{=}\frac{1}{2\pi}\int_{\partial\mathbb{D}}\left|\log\left|\frac{f(\zeta)}{P(\zeta)}\right|\right|\frac{1-|z|^2}{|\zeta-z|^2}|d\zeta|.$$

Then there exists a number A_n depending only on n such that if $M_f(z)\geqslant A_n\rho^\alpha$, then in the region

$$\Omega(z)=\left\{\zeta\in\mathbb{D}:\frac{\rho}{2}\leqslant 1-|\zeta|\leqslant(n+2)\rho,\quad\left|\arg\frac{\zeta}{z}\right|\leqslant\pi\rho\right\}.$$

We have an estimate

$$|f(\zeta)|\leq C_{n1}e^{-C_{n2}B}M_f(z). \tag{1.1.1}$$

PROOF. Let C_n be such a number that

$$|f(\zeta)-P(\zeta)|\leq C_n|\zeta-z_0|^{\alpha},\quad \zeta\in\partial\mathbb{D}.$$

Then if we take A_n to be equal to. say $(4\pi)^{n+2}C_n$, then for $M_f(z)\geqslant A_n\rho^{\alpha}$ we obtain

$$\frac{1}{2}M_f(z)\leqslant M_P(z)\leqslant 2M_f(z). \tag{1.1.2}$$

We also notice that Lemmas 1.4 and 1.1 yield

$$|Q(z^*)|\underset{n}{\asymp}M_P(z),\quad z^*\in\Omega(z). \tag{1.1.3}$$

Finally Lemma 1.2, (1.1.2) and (1.1.3) imply

$$|f(z^*)|\leqslant \exp\frac{1}{2\pi}\int\limits_{\partial\mathbb{D}}\log|f(\zeta)|\frac{1-|z^*|^2}{|\zeta-z^*|^2}|d\zeta|=$$

$$=|Q(z^*)|\exp\frac{1}{2\pi}\int\limits_{\partial\mathbb{D}}\log\left|\frac{f(\zeta)}{P(\zeta)}\right|\frac{1-|z^*|^2}{|\zeta-z^*|^2}|d\zeta|\underset{n}{\asymp}$$

$$\underset{n}{\asymp}M_f(z)\exp\frac{1}{2\pi}\int\limits_{|f(\zeta)|<|P(\zeta)|}\log\left|\frac{f(\zeta)}{P(\zeta)}\right|\frac{1-|z^*|^2}{|\zeta-z^*|^2}|d\zeta|\cdot\exp\frac{1}{2\pi}\int\limits_{|f(\zeta)|\geqslant|P(\zeta)|}\log\left|\frac{f(\zeta)}{P(\zeta)}\right|\cdot\frac{1-|z^*|^2}{|\zeta-z^*|^2}|d\zeta|\leqslant$$

$$\leqslant M_f(z)\exp\frac{C_{n2}}{2\pi}\int\limits_{|f(\zeta)|<|P(\zeta)|}\log\left|\frac{f(\zeta)}{P(\zeta)}\right|\frac{1-|z|^2}{|\zeta-z|^2}\cdot\exp\frac{1}{2\pi}\int\limits_{\partial\mathbb{D}}\log\frac{|P(\zeta)|+C_n|\zeta-z_0|^{\alpha}}{|P(\zeta)|}\cdot\frac{1-|z^*|^2}{|\zeta-z^*|^2}|d\zeta|\leqslant$$

$$\leqslant M_f(z)\cdot\exp(-C_{n2}B)\exp C_{n3}\int\limits_{\partial\mathbb{D}}\log\frac{|P(\zeta)|+C_n|\zeta-z_0|^{\alpha}}{|P(\zeta)|}\cdot\frac{1-|z|^2}{|\zeta-z|^2}|d\zeta|\leqslant C_{n1}e^{-C_{n2}B}M_f(z). \tag{1.1.3}$$ ●

LEMMA 1.6. Let $f\in C^{\alpha}[a,b]$, $n<\alpha<n+1$, $x_0,\dots,x_{n+1}\in[a,b]$, $x_\nu=x_0+\nu h$ and let $|f^{(n)}(x)-f^{(n)}(y)|\leqslant|x-y|^{\alpha-n}, x,y\in[a,b]$. Then

$$\left|\sum_{\nu=0}^{n+1}(-1)^{\nu}C_{n+1}^{\nu}f(x_\nu)\right|\leqslant A_{n0}h^{\alpha}.$$

Proof see [32] , ch.3. ●

We proceed now to the proof of part a9 of Theorem 5. We can assume $|f^{(n)}(\zeta)-f^{(n)}(\xi)|\leqslant|\zeta-\xi|^{\alpha-n}, \zeta,\xi\in\bar{\mathbb{D}}$. Suppose first that $M_f(z)\geqslant 2(A_n+A_{n0})\rho^{\alpha}$ (*) where A_n is taken from Lemma 1.5 and A_{n0} is taken from Lemma 1.6. Let $z^*\in\gamma(z)$ satisfy $z^*_0\in\gamma(z)$,

$$|f(z^*_0)|=M_f(z).$$

Let $z^*_\nu=(1-\nu\rho)z^*_0$, $\nu=1,\dots,n+1$. Lemma 1.6 implies

$$\Big|\sum_{\nu=0}^{n+1}(-1)^\nu C^\nu_{n+1}f(z^*_\nu)\Big|\leqslant A_{n0}\rho^\alpha,$$

and this together with Lemma 1.5 yields

$$M_f(z)=|f(z^*_0)|\leqslant\Big|\sum_{\nu=0}^{n+1}(-1)^\nu C^\nu_{n+1}f(z^*_\nu)\Big|+$$

$$+\Big|\sum_{\nu=1}^{n+1}(-1)^\nu C^\nu_{n+1}f(z^*_\nu)\Big|\leqslant A_{n0}\rho^\alpha+2^{n+1}C_{n1}e^{-C_{n2}I_\rho(f;z)}\cdot M_f(z), \qquad (1.1.4)$$

because $z^*_\nu\in\Omega(z)$, $1\leqslant\nu\leqslant n+1$ we deduce from (1.1.4) that

$$2^{n+1}C_{n1}e^{-C_{n2}I_\rho(f;z)}\geqslant\frac{1}{2}, \qquad (1.1.5)$$

because otherwise we would have $M_f(z)<2A_{n0}\rho^\alpha$ which contradicts our assumption. Now (1.1.5) yields $I_\rho(f;z)\leqslant C_{n3}$ and since by Lemma 1.3 we have $|I_\rho(f;z)-\frac{1}{2\pi}I_M(f;z)|\leqslant_n 1$ it follows that

$$I_M(f;z)\leqslant C_{n4}. \qquad (1.1.6)$$

LEMMA 1.7. Let $f\in C^\alpha(\partial\mathbb{D})$ and let arcs $I\subset\mathcal{J}\subset\partial\mathbb{D}$ satisfy $I\subset\mathcal{J}\subset\partial\mathbb{D}$, $|\mathcal{J}|\leqslant A_1|I|$, $M_f(I)\geqslant A_2|I|^\alpha$.

Then

$$M_f(\mathcal{J}) \underset{A_1, A_2, \alpha}{\preccurlyeq} M_f(I).$$

PROOF. It is an easy consequence of Lemma 1.1. ●

We suppose now that $M_f(z) \geqslant \rho^\alpha$. Let $z^* \in \gamma(z)$ satisfy $z^* \in \gamma(z)$, $|f(z^*)| = M_f(z)$, we put $A_{n1} = 2(A_n + A_{n0})$, $\tilde{\rho} = \rho / A_{n1}^{1/\alpha}$, $\tilde{z} = (1-\tilde{\rho})z^*$. Then

$$M_f(\tilde{z}) = M_f(\gamma(\tilde{z})) \geqslant |f(z^*)| \geqslant \rho^\alpha = A_{n1}\tilde{\rho}^\alpha,$$

and we can apply the estimate (1.1.6) to $\check{z}$. Lemma 1.7 implies $M_f(\tilde{z}) \asymp M_f(z)$. Hence

$$\int_{\partial\mathbb{D}} |\log| \frac{f(\zeta)}{M_f(z)} \| \frac{1-|z|^2}{|\zeta - z|^2} |d\zeta| \underset{n}{\preccurlyeq} \int_{\partial\mathbb{D}} |\log| \frac{f(\zeta)}{M_f(z)} \| \frac{1-|\tilde{z}|^2}{|\zeta - \tilde{z}|^2} |d\zeta| \leqslant$$

$$\leqslant \int_{\partial\mathbb{D}} |\log| \frac{f(\zeta)}{M_f(\tilde{z})} \| \frac{1-|\tilde{z}|^2}{|\zeta - \tilde{z}|^2} |d\zeta| + 2\pi \log \frac{M_f(\tilde{z})}{M_f(z)} \underset{n}{\preccurlyeq} 1.$$

which completes the proof of part a). ●

1.2. Some technicalities.

Let P_1, P_2 be polynomial, $Q_j = {}_eP_j$, $j = 1,2$. Then Q_1, Q_2 are polynomials too and we have

$$\varphi(\zeta) = \frac{Q_1(\zeta)}{Q_2(\zeta)} \exp \frac{1}{2\pi} \int_{\partial\mathbb{D}} \log \left| \frac{P_2(t)}{P_1(t)} \right| \frac{t+\zeta}{t-\zeta} |dt| \equiv 1. \tag{1.2.1}$$

It turns out that if the zeros of P_1 and P_2 are for apart from z and P_1 and P_2 are close in a sense while the integral in (1.2.1) is taken only along an arc placed near the zeros of the P_j's , then the function φ in (1.2.1) approximates 1 and their derivatives approximate 0 .

LEMMA 1.8. a) Let L be an annulus $D(z, x, X)$, where

$x \geqslant 2\rho$, $\frac{4}{3} \leqslant \frac{X}{x} \leqslant \frac{5}{2}$, P_1 and P_2 be polynomials $\deg P_j = n$, $j=1,2$, and let all the zeros of P_1 lie in the annulus

$$L_0 = D(z, q_n'' x, q_n' X), \quad q_n' = 1 - \frac{1}{4(n+1)^2},$$

$$q_n'' = 1 + \frac{1}{4(n+1)^2}, \quad I = L \cap \partial\mathbb{D}, \quad Q_j = {}_e P_j, \quad j = 1,2.$$

We put

$$\varphi(t) \stackrel{def}{=\!=} \frac{Q_1(t)}{Q_2(t)} \exp \frac{1}{2\pi} \int_I \log \left| \frac{P_2(\tau)}{P_1(\tau)} \right| \frac{\tau + t}{\tau - t} |d\tau|, \quad t \in \mathbb{D}.$$

Then there exists a number $\sigma_n > 0$ depending only on n such that if $0 \leqslant \sigma \leqslant \sigma_n$ and

$$|P_1(\zeta) - P_2(\zeta)| \leqslant \sigma \max_I |P_1(t)|, \quad \zeta \in I \tag{1.2.2}$$

then

$$|\varphi(z)| \leqslant C_{0n}$$
$$|\varphi^{(\nu)}(z)| \leqslant C_{\nu n} \sigma X^{-\nu}, \quad \nu = 1, 2, \ldots \tag{1.2.3}$$

where C_{ν_n}, $\nu = 0, 1, \ldots$ depend only on ν and n ;

b) the statement of part a) is true if L is a disc $D(z, X)$, $2\rho \leqslant X \leqslant 4\rho$, all the zeros of P_1 lie in the disc $D(z, q_n' X)$, $I = L \cap \partial\mathbb{D}$ and (1.2.2), (1.2.3) are rewritten literally.

PROOF. We prove only part a) because part b) is simple. Let

$$\gamma_1 = \partial\mathbb{D} \cap D(z, x), \quad \gamma_2 = \partial\mathbb{D} \setminus D(z, X).$$

Then

$$\varphi(t) = \exp \frac{-1}{4\pi} \int_{\gamma_1} \log \left| \frac{P_2(\tau)}{P_1(\tau)} \right|^2 \frac{\tau + t}{\tau - t} |d\tau| \cdot \exp \frac{-1}{2\pi} \int_{\gamma_2} \log \left| \frac{P_2(\tau)}{P_1(\tau)} \right| \frac{\tau + t}{\tau - t} |d\tau| =$$

$$\stackrel{def}{=\!=} \varphi_1(t) \cdot \varphi_2(t).$$

We put $\psi_j = -\log \varphi_j$, $j=1,2$ and change the formula for the ψ_1.

Let

$$P_j(\tau) = \sum_{\nu=0}^{n} a_\nu^{(j)} \tau^\nu,$$

$$P_j^*(\tau) = \sum_{\nu=0}^{n} \bar{a}_\nu^{(j)} \tau^{n-\nu}, \quad j=1,2.$$

For $\tau \in \partial\mathbb{D}$ we have

$$|P_j(\tau)|^2 = P_j(\tau)\cdot\overline{P_j(\tau)} = \sum_0^n a_\nu^{(j)}\tau^\nu \cdot \sum_0^n \bar{a}_\mu^{(j)} \tau^{-\mu} = \tau^{-n} P_j(\tau) P_j^*(\tau),$$

and hence for $\tau \in \partial\mathbb{D}$ we have

$$\left|\frac{P_2(\tau)}{P_1(\tau)}\right|^2 = \frac{P_2(\tau)}{P_1(\tau)} \cdot \frac{P_2^*(\tau)}{P_1^*(\tau)}.$$

Before writing a new formula for ψ_1 we point out important estimates for P_j and P_j^*. If $\tau \in I$ then

$$|P_2^*(\tau) - P_1^*(\tau)| = |P_2(\tau) - P_1(\tau)| \leq \sigma \max_I |P_1(t)| = \sigma \max_I |P_1^*(t)|. \tag{1.2.4}$$

Let L^* be a region symmetric to L with respect to $\partial\mathbb{D}$. The zeros of P_1^* lie in L_0^* symmetric to L_0 with respect to $\partial\mathbb{D}$. Using lemmas 1.1 and the assumptions of Lemma 1.8 we obtain

$$|P_1(\zeta)| \geq C_{n1} \max_I |P_1|, \quad \zeta \in \partial L, \tag{1.2.5}$$

$$|P_1^*(\zeta)| \geq C_{n1} \max_I |P_1^*|, \quad \zeta \in \partial L^*, \tag{1.2.6}$$

$$\max_{\partial L} |P_2(\zeta) - P_1(\zeta)| \leq C_{n2} \max_I |P_2(\zeta) - P_1(\zeta)|, \tag{1.2.7}$$

$$\max_{\partial L^*} |P_2^*(\zeta)-P_1^*(\zeta)| \leqslant C_{n2} \max_{I} |P_2^*(\zeta)-P_1^*(\zeta)|. \qquad (1.2.8)$$

Using (1.2.4) - (1.2.8) we obtain

$$\left| \frac{P_2(\zeta)-P_1(\zeta)}{P_1(\zeta)} \right| \leqslant C_{n3}\sigma, \quad \zeta \in \partial L, \qquad (1.2.9)$$

$$\left| \frac{P_2^*(\zeta)\, P_1^*(\zeta)}{P_1^*(\zeta)} \right| \leqslant C_{n3}\sigma, \quad \zeta \in \partial L^*. \qquad (1.2.10)$$

By the maximum principle the inequalities (1.2.9) and (1.2.10) hold in $\mathbb{C} \setminus L$ and $\mathbb{C} \setminus L^*$. Hence if $0 \leqslant \sigma \leqslant \sigma_n$, where $C_{n3}\sigma_n = \frac{1}{2}$ then (1.2.9), (1.2.10) yield

$$\left|\log \frac{P_2(\zeta)}{P_1(\zeta)}\right| \leqslant C_{n4}\sigma, \quad \zeta \in \mathbb{C} \setminus L, \qquad (1.2.11)$$

$$\left|\log \frac{P_2^*(\zeta)}{P_1^*(\zeta)}\right| \leqslant C_{n4}\sigma, \quad \zeta \in \mathbb{C} \setminus L^*. \qquad (1.2.12)$$

We write now a convenient formula for φ_1. Let $\tilde{\gamma}_1$ be the arc of the circle which has the same endpoints as γ_1 lies outside $\mathbb{D} \cup L \cup L^*$ and constitute with the $\partial \mathbb{D}$ the angle $\geqslant \alpha_0 > 0$, α_0 being an absolute constant (for example $\alpha_0 = 0{,}0001\pi$). Due to geometrical properties of L and L^* we can find such an arc. Then for any point $\zeta \in \tilde{\gamma}_1$ $|\zeta - z| \asymp X$ and

$$\varphi_1(t) = \frac{1}{4\pi i} \int_{\tilde{\gamma}_1} \left(\log \frac{P_2(\tau)}{P_1(\tau)} + \log \frac{P_2^*(\tau)}{P_1^*(\tau)} \right) \frac{\tau + t}{\tau - t} \, \frac{d\tau}{\tau}, \quad t \in \mathbb{D}.$$

The inequalities (1.2.11) and (1.2.12) hold on $\tilde{\gamma}_1$ and γ_2. Hence

$$|\mathrm{Re}\,\varphi_1(z)| = \left|\mathrm{Re}\,\frac{1}{4\pi i}\int_{\tilde{\gamma}_1}\left(\log\frac{P_2(\tau)}{P_1(\tau)} + \log\frac{P_2^*(\tau)}{P_1^*(\tau)}\right)\frac{\tau+z}{\tau-z}\,\frac{d\tau}{\tau}\right| \leqslant$$

$$\leqslant C_0\cdot C_{n4}\,\sigma\int_{\tilde{\gamma}_1}\frac{|d\tau|}{|\tau-z|} \leqslant C_{n5}\,\sigma; \tag{1.2.13}$$

$$|\mathrm{Re}\,\varphi_2(z)| \leqslant 2C_{n4}\,\sigma\cdot\frac{1}{2\pi}\int_{\gamma_2}\frac{1-|z|^2}{|\tau-z|^2}|d\tau| \leqslant C_{n5}\,\sigma; \tag{1.2.14}$$

$$|\varphi_1^{(\nu)}(z)| = \left|\frac{\nu!}{2\pi i}\int_{\tilde{\gamma}_1}\left(\log\frac{P_2(\tau)}{P_1(\tau)} + \log\frac{P_2^*(\tau)}{P_1^*(\tau)}\right)\frac{d\tau}{(\tau-z)^{\nu+1}}\right| \leqslant$$

$$\leqslant C_{n4}\,C_\nu\,\sigma\cdot\int_{\tilde{\gamma}_1}\frac{|d\tau|}{|\tau-z|^{\nu+1}} \leqslant C_{\nu n}\,\sigma X^{-\nu}, \quad \nu\geqslant 1; \tag{1.2.15}$$

$$|\varphi_2^{(\nu)}(z)| = \left|\frac{\nu!}{\pi i}\int_{\gamma_2}\log\left|\frac{P_2(\tau)}{P_1(\tau)}\right|\frac{|d\tau|}{(\tau-z)^{\nu+1}}\right| \leqslant$$

$$\leqslant C_{n4}\,C_\nu\,\sigma\int_{\gamma_2}\frac{|d\tau|}{|\tau-z|^{\nu+1}} \leqslant C_{\nu n}\,\sigma X^{-\nu}, \quad \nu\geqslant 1. \tag{1.2.16}$$

Differentiating the function $\varphi(t) = e^{-\varphi_1(t)}\cdot e^{-\varphi_2(t)}$ and putting $t=z$ we obtain from (1.2.14) - (1.2.16) the statement of Lemma 1.8. ●

1.3. The first crucial estimate.

LEMMA 1.9. a) Let L be an annulus $D(z,x,X)$, $2\rho\leqslant x<X<\frac{1}{2}$, $\frac{4}{3}\leqslant\frac{X}{x}\leqslant\frac{5}{2}$ the points $\zeta_1,\dots,\zeta_m$ lie in the annulus $L_0 = D(z, q_n'' x, q_n' X)$, $n\geqslant m$, $P(t) \overset{def}{=} \prod_{\nu=1}^{m}(t-\zeta_\nu)$, $I = L\cap\partial\mathbb{D}$, $J = D(z,X)\cap\partial\mathbb{D}$, $Q = {}_eP$. Let the function $v\in C^m(I)$ and

$$|v^{(\nu)}(\zeta)| \leq \sigma X^{m-\nu}, \quad \nu = 0,1,\dots,m, \quad \zeta \in I;$$

(1.3.0)

$$\varphi(\zeta) \stackrel{def}{=} Q(\zeta) \exp \frac{1}{2\pi} \int_I \log\left|1 - \frac{v(t)}{P(t)}\right| \frac{t+\zeta}{t-\zeta} |dt|, \quad \zeta \in D.$$

Then there exists $\sigma_{mn} > 0$ such that for $0 \leq \sigma \leq \sigma_{mn}$ we have

$$|\varphi^{(\nu)}(z)| \leq C_{mn\nu} X^{m-\nu}, \quad \nu = 0,1,\dots,m, \tag{1.3.1}$$

$$|\varphi^{(\nu)}(z)| \leq C_{mn\nu} \sigma X^{m-\nu}, \quad \nu = m+1, m+2, \dots \tag{1.3.2}$$

b) The statements (1.3.1) and (1.3.2) of part a) hold true if we put L to be a disc $D(z,X)$, $2\rho \leq X \leq 4\rho$, L_0 to be a disc $D(z, q_n^1 X)$ and if $I = L \cap \partial D$.

PROOF. We only consider the most complicated case a).

We use the induction in m. If $m=0$ we put $P \equiv 1$. Then if

$$|v(\zeta)| \leq \sigma \leq \sigma_{0n} = \frac{1}{2},$$

then

$$|\log|1+\sigma|| \lesssim \sigma,$$

and if $\psi = \log \varphi$ then

$$|\psi(z)| \lesssim \sigma, \tag{1.3.3}$$

$$|\psi^{(\nu)}(z)| \underset{\nu}{\lesssim} \sigma X^{-\nu}, \quad \nu = 1,2,\dots, \tag{1.3.4}$$

and we obtain (1.3.1), (1.3.2) differentiating e^{ψ} and the substitution (1.3.3) and (1.3.4).

We proceede now to the inductive step. Suppose that all numbers $\sigma_{kn},\ 0 \leqslant k \leqslant m,\quad n \geqslant k$ have already been choosen. Let

$$\zeta_1, \dots, \zeta_{m+1} \in L_0,\quad v \in C^{m+1}(I),\quad n \geqslant m+1,$$

$$P(t) = \prod_{\nu=1}^{m+1} (t - \zeta_\nu),\quad f \overset{def}{=} P + v.$$

We extend v to the whole arc $\mathcal{J}$ in such a way that (1.3.0) is satisfied with the constant $\tilde{C}_{m+1}\sigma$ instead of σ. We put for $x \in \mathcal{J}$

$$P_x(\zeta) \overset{def}{=} \sum_{\nu=0}^{m+1} \frac{f^{(\nu)}(x)}{\nu!} (\zeta - x)^\nu. \tag{1.3.5}$$

We need some properties of P_x. Let a_x be the senior coefficient of P_x. We have

$$a_x = \frac{f^{(m+1)}(x)}{(m+1)!} = 1 + \frac{v^{(m+1)}(x)}{(m+1)!} \in C(\mathcal{J}),$$

if $|v^{(m+1)}(x)| \leqslant \frac{1}{2}$ then $|a_x| \geqslant \frac{1}{2}$; (1.3.5')

$$|a_x - 1| \underset{m+1}{\leqslant \sigma},\quad x \in \mathcal{J}; \tag{1.3.6'}$$

if $\xi \in \mathcal{I}$ then

$$|P(\zeta) - P_x(\zeta)| \leqslant |v(\zeta)| + |f(\zeta) - P_x(\zeta)| =$$

$$= |v(\zeta)| + \frac{1}{m!} \left| \int_x^\zeta (\zeta - t)^m (f^{(m+1)}(t) - f^{(m+1)}(x))\,dt \right| =$$

$$= |v(\zeta)| + \frac{1}{m!} \int_x^\zeta (\zeta - t)^m (v^{(m+1)}(t) - v^{(m+1)}(x))\,dt \leqslant$$

$$\leqslant C^0_{m+1} \sigma X^{m+1}, \tag{1.3.6}$$

Using (1.3.5) and Lemma 1.1 we obtain

$$|P(\zeta)-\frac{1}{a_x}P_x(\zeta)|\leqslant C_{n2}\sigma \max_I |P|, \quad \zeta\in\mathcal{J},\ x\in\mathcal{J}. \tag{1.3.7}$$

Since all the zeros of P lie in L_0 we find, putting $\tilde{L}=D(z,q''_{n+1}x,q'_{n+1}X)$, that

$$|P(\zeta)|\geqslant C_{m+1,n}X^{m+1}, \quad \zeta\in\partial\tilde{L}, \tag{1.3.8}$$

and then if

$$C^{0}_{m+1}\sigma\leqslant\frac{1}{2}C_{m+1,n}, \tag{1.3.9}$$

then (1.3.6) and (1.3.8) yield

$$|P_x(\zeta)-P(\zeta)|\leqslant\frac{1}{2}C_{m+1,n}X^{m+1}\leqslant\frac{1}{2}|P(\zeta)|,\quad \zeta\in\partial\tilde{L},\ x\in\mathcal{J},$$

The Rouchet theorem implies that all the zeros of P_x lie in $\tilde{L}$ for any $x\in\mathcal{J}$. Since all coefficients of P_x depend continuously on x we may number their zeros $\zeta_1(x),\dots,\zeta_{n+1}(x)$, $\zeta_\nu(x)\in\tilde{L}$ in such a way that $\arg\zeta_\nu(x)$ depends continuously on $x\in\mathcal{J}$ and consequently on $\theta=\arg x$, where

$$\theta\in[\Theta_1,\Theta_2],\ \Theta_1<\Theta_2,\ \{e^{i\Theta_j}\}=\partial\mathbb{D}\cap D(z,q'_{n+1}X),\ j=1,2.$$

The function $G(\theta)=\arg\zeta_1(\theta)-\theta$, $\theta\in[\Theta_1,\Theta_2]$ is continuous on $[\Theta_1,\Theta_2]$, $G(\Theta_1)\geqslant 0$, $G(\Theta_2)\leqslant 0$, i.e. we may find $\theta_0\in[\Theta_1,\Theta_2]$ for which $G(\theta_0)=0$. So there exists $x_1\in D(z,q'_{n+1}X)\cap\partial\mathbb{D}$, for which $\arg\zeta_1(x_1)=\arg x_1$.

We fix the point x_1 and let

$$a=a_{x_1}=\frac{f^{(m+1)}(x_1)}{(m+1)!},\qquad v_{x_1}=f-P_{x_1},$$

$$V(t)=\frac{v_{x_1}(t)}{a(t-\zeta_1(x_1))}.$$

We check that $V(t)$ satisfies the conditions (1.3.0) with the replacement of σ by $\tilde{C}_{m+1}\sigma$ for the ν-th derivatives,

Since P_{x_1} is a Taylor polynomial of f at x_1, we have

$$v_{x_1}^{(\nu)}(x_1)=(f-P_{x_1})^{(\nu)}(x_1)=0, \quad \nu=0,\dots,m+1,$$

hence

$$v_{x_1}(t)=\frac{1}{m!}\int_{x_1}^{t}(t-\tau)^m(f^{(m+1)}(\tau)-f^{(m+1)}(x_1))\,d\tau=$$

$$=\frac{1}{m!}\int_{x_1}^{t}(t-\tau)^m(v^{(m+1)}(\tau)-v^{(m+1)}(x_1))\,d\tau$$

i.e.

$$V(t)=\frac{1}{am!}\cdot\frac{1}{t-\zeta_1(x_1)}\int_{x_1}^{t}(t-\tau)^m(v^{(m+1)}(\tau)-v^{(m+1)}(x_1))\,d\tau$$

and consequently, to estimate $V^{(\nu)}$, $0\leqslant\nu\leqslant m$ it is sufficient to estimate the terms of the form

$$T_{\mu\nu}=\frac{1}{(t-\zeta_1)(x))^{\mu+1}}\int_{x_1}^{t}(t-\tau)^{m-n+\mu}(v^{(m+1)}(\tau)-v^{(m+1)}(x_1))\,d\tau, \quad 0\leqslant\mu\leqslant\nu.$$

Since $m-\nu+\mu\geqslant 0$ and $|t-\zeta_1(x_1)|\geqslant|t-x_1|$, $t\in J$, (we recall that $\arg\zeta_1(x_1)=\arg x_1$) we have

$$|T_{\mu\nu}|\underset{m}{\leqslant}\frac{1}{|t-x_1|^{\mu+1}}\cdot|t-x_1|^{m-\nu+\mu+1}\sigma,$$

i.e.

$$|V^{(\nu)}(t)|\leqslant\tilde{C}_{m+1,0}\,\sigma X^{m-\nu}, \quad 0\leqslant\nu\leqslant m. \tag{1.3.10}$$

Let

$$l(t)=a(t-\zeta_1(x_1)), \qquad P_0(t)=P_{x_1}(t)/l(t),$$

$$Q_{x_1} = {}_eP_{x_1}, \quad Q_0 = {}_eP_0, \quad \lambda = {}_e\ell; \quad Q_{x_1} = \lambda Q_0 .$$

We may now write as follows:

$$\varphi(\zeta) = \left[|a| \frac{Q(\zeta)}{Q_{x_1}(\zeta)} \cdot \exp \frac{1}{2\pi} \int_I \log \left| \frac{P_{x_1}(t)}{P(t)} \right| \frac{t+\zeta}{t-\zeta} |dt| \right] \times$$

$$\times \left[\frac{1}{|a|} Q_{x_1}(\zeta) \cdot \exp \frac{1}{2\pi} \int_I \log \left| 1 + \frac{v_{x_1}(t)}{P_{x_1}(t)} \right| \frac{t+\zeta}{t-\zeta} |dt| \right] =$$

$$= \left[|a| \frac{Q(\zeta)}{Q_{x_1}(\zeta)} \cdot \exp \frac{1}{2\pi} \int_I \log \left| \frac{P_{x_1}(t)}{P(t)} \right| \frac{t+\zeta}{t-\zeta} |dt| \right] \times$$

$$\times \left[Q_0(\zeta) \cdot \exp \frac{1}{2\pi} \int_I \log \left| 1 + \frac{V(t)}{P_0(t)} \right| \frac{t+\zeta}{t-\zeta} |dt \right] \cdot \frac{1}{|a|} \lambda(\zeta) \overset{def}{=\!=} \Phi_{x_1}(\zeta) \varphi_{x_1}(\zeta) \lambda(\zeta),$$

$$\Phi_{x_1}(\zeta) \overset{def}{=\!=} \frac{Q(\zeta)}{Q_{x_1}(\zeta)} \cdot \exp \frac{1}{2\pi} \int_I \log \left| \frac{P_{x_1}(t)}{P(t)} \right| \frac{t+\zeta}{t-\zeta} |dt| .$$

The relations (1.3.10) permit us to apply the inductive assumption to the factor φ_{x_1} under suitable choice of σ . To the factor Φ_{x_1} we can apply the results of 1.2 for small σ . The factor $\lambda(\zeta)$ has the form $c(\zeta - \zeta_1^*)$ where $|c| \asymp 1$, $\zeta_1^* \in \tilde{L} \cup \tilde{L}^*$, $\tilde{L}^*$ is symmetric to $\tilde{L}$ with respect to $\partial\mathbb{D}$ and we have

$$|\lambda(z)| \asymp X, \quad |\lambda'(z)| \asymp 1, \quad \lambda^{(\nu)}(z) = 0, \quad \nu \geqslant 2.$$

We fix the value of $\sigma_{m+1,n}$. We have

$$|\varphi(z)| \underset{m+1,n}{\lesssim} 1 \cdot X^m \cdot X = X^{m+1};$$

$$\varphi^{(\nu)}(z) = \lambda(z) (\Phi_{x_1} \varphi_{x_1})^{(\nu)}(z) + \nu \lambda'(z) (\Phi_{x_1} \varphi_{x_1})^{(\nu-1)}(z), \quad \nu \geqslant 1,$$

It follows that we should estimate the terms of the form $\lambda\Phi_{x_1}^{(\mu)}\varphi_{x_1}^{(\nu-\mu)}$, $\mu\leqslant\nu$ and $\lambda'\Phi_{x_1}^{(\mu)}\varphi_{x_1}^{(\nu-\mu-1)}$, $\mu\leqslant\nu-1$. Suppose first $1\leqslant\nu\leqslant m+1$. Then because of $|\Phi_{x_1}^{(\mu)}(z)|\underset{n,\mu}{\preccurlyeq} X^{-\mu}$ we obtain

$$|\lambda(z)\Phi_{x_1}^{(\mu)}(z)\varphi_{x_1}^{(\nu-\mu)}(z)|\underset{m+1,n,\nu}{\preccurlyeq} X\cdot X^{-\mu}\cdot X^{m-(\nu-\mu)}=X^{m+1-\nu},$$

$$|\lambda'(z)\Phi_{x_1}^{(\mu)}(z)\varphi_{x_1}^{(\nu-\mu-1)}(z)|\underset{m+1,n,\nu}{\preccurlyeq} 1\cdot X^{-\mu}\cdot X^{m-(\nu-\mu-1)}=X^{m+1-\nu}.$$

Suppose now $\nu>m+1$. If $\mu>0$ then

$$|\lambda\Phi_{x_1}^{(\mu)}\varphi_{x_1}^{(\nu-\mu)}|\underset{m+1,n,\nu}{\preccurlyeq} X\cdot\sigma X^{-\mu}\cdot X^{m-(\nu-\mu)}=\sigma X^{m+1-\nu},$$

$$|\lambda'\Phi_{x_1}^{(\mu)}\varphi_{x_1}^{(\nu-\mu-1)}|\underset{m+1,n,\nu}{\preccurlyeq} 1\cdot\sigma X^{-\mu}\cdot X^{m-(\nu-\mu-1)}=\sigma X^{m+1-\nu}.$$

If $\mu=0$ then $\nu-1>m$ and

$$|\lambda\Phi_{x_1}\varphi_{x_1}^{(\nu)}|\underset{m+1,n,\nu}{\preccurlyeq} X\cdot 1\cdot\sigma X^{m-\nu}=\sigma X^{m+1-\nu},$$

$$|\lambda'\Phi_{x_1}\varphi_{x_1}^{(\nu-1)}|\underset{m+1,n,\nu}{\preccurlyeq} 1\cdot 1\cdot\sigma X^{m-(\nu+1)}=\sigma X^{m+1-\nu}.$$

Hence to finish the proof it is sufficient to choose $\sigma_{m+1,n}$, $n\geqslant m+1$ in such a way that we could use (1.3.5'), (1.3.9), 1.2 and the inductive assumption. We write down these inequalities

$$\tilde{C}_{m+1}\sigma_{m+1,n}\leqslant\frac{1}{2} \qquad \text{for (1.3.5')} \qquad (1.3.11)$$

$$C^{o}_{m+1}\sigma_{m+1,n}\leqslant\frac{1}{2}C_{m+1,n} \qquad \text{for (1.3.9)} \qquad (1.3.12)$$

$$\tilde{C}_{m+1,0}\sigma_{m+1,n}\leqslant\sigma \qquad \text{for (1.3.0)} \qquad (1.3.13)$$

$$C_{n2}\sigma_{m+1,n} \leq \sigma_{n+1} \text{ from (1.2)} \tag{1.3.14}$$

In (1.3.13) and (1.3.14) the value $n+1$ is taken instead of n since we used the annulus

$$\tilde{L} = D(z, q''_{n+1}x, q'_{n+1}X),$$

corresponding $n+1$;

σ_{n+1} from (1.2) means the value σ_{n+1} in 1.2.

A choice of $\sigma_{m+1,n}$ satisfying (1.3.11) - (1.3.14) completes the proof of Lemma 1.9. ●

1.4. The second crucial estimate

LEMMA 1.10. a) Let $L = D(z, X_1, X_0)$ where $2\rho \leq X_1 < X_0 \leq \frac{1}{2}$, $X_0 \geq \frac{4}{3}X_1$, $I = L \cap \partial\mathbb{D}$.

Let P be a polynomial, $\deg P \leq n$, without zeros in the annulus $L^0 = D(z, q'_n X_1, q''_n X_0)$, $M_j = \max\limits_{D(z,X_j)} |P|$, $j = 0,1$.

Let $v \in C(I)$ be a function such that

$$|v(t)| \leq |t - z_0|^{\alpha}, \quad n < \alpha < n+1,$$

$$\Phi(\zeta) \overset{def}{=\!=} \exp \frac{1}{2\pi} \int_I \log\left|1 - \frac{v(t)}{P(t)}\right| \left|\frac{t+\zeta}{t-\zeta}\right| |dt|, \quad \zeta \in \mathbb{D}$$

and $\sigma_j = \dfrac{X_j^{\alpha}}{M_j}$, $j = 0,1$. Then there exists a number $\sigma_n^0 > 0$ such that if $\sigma_0 \leq \sigma_n^0$ then

$$|\Phi(z)| \underset{n}{\leqslant} 1, \tag{1.4.1}$$

$$|\Phi^{(\nu)}(z)| \underset{n,\nu}{\leqslant} \sigma_0 X_0^{-\nu} + \sigma_1 X_1^{-\nu}, \quad \nu \geqslant 1; \tag{1.4.2}$$

b) (1.4.1) and (1.4.2) hold if $L = D(z, X_0)$, $X_0 \geqslant 2\rho$ and $L^0 = D(z, q_n'' X_0)$ and in the definition of M_1 and σ_1 we put in (1.4.2) $X_1 = \rho$.

PROOF. We prove only the part a).

We suppose that the polynomial $P(t) = a \prod\limits_{|\zeta_\mu - z| > X_0} (t - \zeta_\mu) \cdot \prod\limits_{|\zeta_\nu - z| < X_1} (t - \zeta_\nu)$ has $m \geqslant 0$ zeros in the disc $\mathcal{D}(z, X_1)$ and put $Q = |a| \prod\limits_{|\zeta_\mu - z| > X_0} |z - \zeta_\mu|$. Then the assumptions made about P yield

$$|P(t)| \asymp Q |t - z|^m, \quad t \in D(z, X_1, X_0).$$

Hence we obtain

$$\left|\frac{v(t)}{P(t)}\right| \leqslant C_{n1} \frac{|t-z|}{Q} \leqslant C_{n1} \frac{X_0^{d-m}}{Q} \leqslant C_{n1} \cdot C_{n2} \frac{X_0^d}{M_0} \leqslant \frac{1}{2} \tag{1.4.3}$$

if $t \in I$, $\sigma_0 \leqslant \sigma_n^0$, $\sigma_n^0 \cdot C_{n1} C_{n2} = \frac{1}{2}$ and then

$$|\Phi(z)| = \exp \frac{1}{2\pi} \int_I \log \left|1 + \frac{v(t)}{P(t)}\right| \cdot \frac{1 - |z|^2}{|t - z|^2} |dt| \leqslant 1.$$

If we put $\varphi \overset{\text{def}}{=} \log \Phi$ then the first inequality in (1.4.3) for small σ_n^0 and $\sigma \leqslant \sigma_n^0$ implies

$$|\varphi^{(\nu)}(z)| \underset{\nu,n}{\leqslant} \int_I \frac{|t-z|^{d-m}}{Q} \frac{|dt|}{|t-z|^{\nu+1}} \underset{\nu,n}{\leqslant} \frac{X_0^{d-m-\nu}}{Q} + \frac{X_1^{d-m-\nu}}{Q} \underset{\nu,n}{\asymp} \frac{X_0^{d-\nu}}{M_0} + \frac{X_1^{d-\nu}}{M_1}.$$

But now we have

$$\Phi^{(\nu)}(z) = e^{\varphi(z)} \sum_{(r_1, \ldots, r_j)} C_\nu^{r_1, \ldots, r_j} \varphi^{(r_1)}(z) \ldots \varphi^{(r_j)}(z),$$

where $r_k > 0$, $\sum\limits_1^j r_k = \nu$. We estimate the term $t_{r_1, \ldots, r_j}$:

$$t_{r_1,\dots,r_j} \overset{def}{=} \varphi^{(r_1)}(z)\dots\varphi^{(r_j)}(z).$$

We suppose first $\alpha - m - \nu > 0$. Then $\alpha - m - r_K > 0$, $X_0^{\alpha-m-r_K} > X_1^{\alpha-m-r_K}$ and hence

$$t_{r_1,\dots,r_j} \underset{n,\nu}{\lesssim} \prod_{K=1}^{j}\left(\frac{1}{Q}X_0^{\alpha-m-r_K}\right) = \frac{1}{Q}X_0^{\alpha-m-\nu}\cdot\frac{1}{Q^{j-1}}X_0^{(\alpha-m)(j-1)} \underset{n,\nu}{\lesssim}$$

$$\underset{n,\nu}{\lesssim} \frac{1}{Q}X_0^{\alpha-m-\nu} \asymp \frac{X_0^{\alpha-\nu}}{M_0}. \tag{1.4.4}$$

Suppose now that $\alpha - m - \nu < 0$. If $\alpha - m - r_K > 0$ for all K, $1 \le K \le j$, then we argue as in (1.4.4). So we can assume that $\alpha - m - r_K > 0$, $1 \le K \le S$, $\alpha - m - r_K < 0$, $S+1 \le K \le j$. Then we have

$$t_{r_1,\dots,r_j} \underset{n,\nu}{\lesssim} \prod_{K=1}^{S}\left(\frac{1}{Q}X_0^{\alpha-m-r_K}\right)\cdot\prod_{K=S+1}^{j}\left(\frac{1}{Q}X_1^{\alpha-m-r_K}\right) =$$

$$= \frac{1}{Q}X_1^{\alpha-m-\nu}\cdot\prod_{K=1}^{S}\left[\left(\frac{1}{Q}X_0^{\alpha-m}\right)\left(\frac{X_1}{X_0}\right)^{r_K}\right]\cdot\prod_{K=S+1}^{j}\left(\frac{1}{Q}X_1\right)^{\alpha-m} \underset{n,\nu}{\lesssim}$$

$$\underset{n,\nu}{\lesssim} \frac{1}{Q}X_1^{\alpha-m-\nu} \underset{n}{\asymp} \frac{X_1^{\alpha-\nu}}{M_1}. \tag{1.4.5}$$

Now the lemma follows from (1.4.4) and (1.4.5). ●

1.5. A technical preparation.

LEMMA 1.11. a) Let $L = D(z, x, X)$ where $x \ge 2\rho$, $\frac{4}{3} \le \frac{X}{x} \le \frac{5}{2}$, $\alpha, n < \alpha < n+1$, $I = L \cap \partial\mathbb{D}$, $\mathcal{J} = D(z, X) \cap \partial\mathbb{D}$,

$P(t) = a\prod_{\nu}(t - \zeta_\nu)$ is the Taylor-polynomial of the function $f \in C^\alpha(\mathcal{J})$, $\|f\|_{C^\alpha} \le 1$ at $z_0 = z/|z|$. Suppose that P has $m > 0$ zeros in L and the annuli $D(z, q_n' x, q_n'' x)$, $D(z, q_n' X, q_n'' X)$

do not contain the zeros of P . Put $Q(t)=\prod_{\zeta_\nu \notin L}(t-\zeta_\nu)$,

$$v(\zeta)=\frac{f(\zeta)-P(\zeta)}{Q(\zeta)}, \qquad \zeta\in J, \qquad \sigma=\frac{X^d}{\max_{D(z,X)}|P|}.$$

Then

$$|v^{(\nu)}(\zeta)|\leqslant C_d\,\sigma X^{m-\nu}, \quad \nu=0,\dots,m, \quad \zeta\in I. \tag{1.5.1}$$

b) The statement (1.5.1) of a) remains valid if we put $L=\mathcal{D}(z,X)$ where $2\rho\leqslant X\leqslant 4\rho$, suppose that the polynomial P has no zeros in the annulus $D(z,q_n'X,\ q_n''X)$ and $I=L\cap\partial\mathbb{D}$.

PROOF. We prove only the part a), the proof of b) is the same. Let $\xi=z+\frac{X+x}{2}z_0\in L$, $Q=|Q(\xi)|$ then $|Q(\zeta)|\underset{n}{\asymp}Q$, $\zeta\in L$ because of the location of the zeros of $Q(\xi)$. Further we have $|f^{(\nu)}(\zeta)|\leqslant$ $\leqslant C_d|\zeta-z_0|^{d-\nu}$, $\zeta\in J$, $\nu=0,1,\dots,n$ and the Cauchy-formula yields $|Q^{(\ell)}(\zeta)|\underset{\ell,n}{\preccurlyeq}$ $\underset{\ell,n}{\preccurlyeq} QX^{-\ell}$, $\zeta\in L$, hence for $\mu+\nu=r\leqslant m\leqslant n$ we obtain

$$\left|\left(\frac{1}{Q}\right)^{(\mu)}(\zeta)\right|\underset{n,\mu}{\preccurlyeq}\frac{1}{QX^{\mu}}, \quad \mu=1,2,\dots, \ \zeta\in L,$$

$$\left|(f-P)^{(\mu)}(\zeta)\cdot\left(\frac{1}{Q}\right)^{(\nu)}(\zeta)\right|\underset{d}{\preccurlyeq}\frac{X^{d-r}}{Q}\underset{d,n}{\asymp}\sigma X^{m-r}, \quad \zeta\in I. \tag{1.5.2}$$

Using now (1.5.2) and the Leibnitz-formula we finish the proof of the lemma. ●

LEMMA 1.12. Let Q be a rational function, $\deg Q\leqslant n$, having no zeros and poles in the annulus

$$L=D(z,q_n'x,q_n''X),\ \tfrac{4}{3}x\leqslant X\leqslant\tfrac{5}{2}x, \quad \xi=z+iz_0\frac{X+x}{2}, \quad Q=|Q(\xi)|.$$

Then

$$|Q^{(\nu)}(z)|\underset{\nu,n}{\preccurlyeq}QX^{-\nu}.$$

The proof is a consequence of the Cauchy- formula for $Q^{(\nu)}$. ●

We define now the number $s = s(A;P,z)$ as follows:

$$s(A;P,z) \overset{def}{=} \max\{t: \quad \text{for any } \tau,\ 0\leq\tau\leq t,\ \max_{D(z,\tau)}|P| \geq A\tau^{\alpha}\}$$

where P is a polynomial, $\deg P \leq n$, $n<\alpha<n+1$.

LEMMA 1.13. Let the function $f\in C^{\alpha}(\partial\mathbb{D})$ satisfy the condition b) of Theorem 5, $\|f\|_{C^{\alpha}} \leq 1$, $n<\alpha<n+1$, P is Taylor-polynomial of f at $z_0 = z/|z|$, $A\geq 1$ is a number, the number $s = s(A;P,z)$ is defined above where $4\rho < s < 2$, the number s_0, $\frac{s}{3} \leq s_0 \leq s$ is arbitrary and

$$\varphi_{s_0}(\zeta) \overset{def}{=} \exp\frac{1}{2\pi}\int\limits_{\partial\mathbb{D}\setminus D(z,s_0)} \log\left|\frac{f(t)}{P(t)}\right| \frac{t+\zeta}{t-\zeta}\,|dt| \overset{def}{=} \exp\varphi(\zeta).$$

Then for $A \geq A(\alpha)$ we have

$$|\varphi_{s_0}(z)| \underset{A,n}{\lessapprox} 1 \tag{1.5.3}$$

$$|\varphi_{s_0}^{(\nu)}(z)| \underset{A,n,\nu}{\lessapprox} s_0^{-\nu}, \qquad \nu = 1,2,\ldots \tag{1.5.4}$$

PROOF. We have

$$|\varphi_{s_0}(z)| = \exp\frac{1}{2\pi}\int\limits_{\partial\mathbb{D}\setminus D(z,s_0)} \log\left|\frac{f(t)}{P(t)}\right| \frac{1-|z|^2}{|t-z|^2}\,|dt|$$

then

$$\int\limits_{\partial\mathbb{D}\setminus D(z,s_0)} \log\left|\frac{f(t)}{P(t)}\right| \frac{1-|z|^2}{|t-z|^2}\,|dt| \leq \int\limits_{\partial\mathbb{D}} \left|\log\left|\frac{f(t)}{P(t)}\right|\right| \frac{1-|z|}{|t-z|^2}\,|dt| \leq$$

$$\leq \int\limits_{\partial\mathbb{D}} \left|\log\left|\frac{f(t)}{M_f(z)}\right|\right| \frac{1-|z|^2}{|t-z|^2}\,|dt| + \int\limits_{\partial\mathbb{D}} \left|\log\left|\frac{P(t)}{M_f(t)}\right|\right| \frac{1-|z|^2}{|t-z|^2}\,|dt| \overset{def}{=}$$

$$\overset{def}{=} T_1 + T_2 \underset{A,n}{\lessapprox} 1. \tag{1.5.5}$$

and because $s \geqslant 4\rho$ we obtain

$$M_P(z) = \max_{\gamma(z)} |P| \geqslant C_{\alpha 0} A \rho^{\alpha}.$$

If $A \geqslant A(\alpha) = 2\tilde{C}_\alpha / C_{\alpha 0}$ where $\tilde{C}_\alpha$ is the constant in the estimate

$$|f(t) - P(t)| \leqslant \tilde{C}_\alpha |t - z_0|^{\alpha}, \quad t \in \partial\mathbb{D},$$

then $M_P(z) \asymp M_f(z)$ and therefore we can apply the condition (1) to the integral (T_1) in (1.5.5), and Lemma 1.3 to the integral (T_2) which yields (1.5.5). We now differentiate the function $\varphi_{s_0} = e^{\psi}$ and estimate the term

$$t_{\tau_1,\dots,\tau_j} \overset{def}{=} \psi^{(\tau_1)}(z)\dots\psi^{(\tau_j)}(z), \quad \tau_1 + \dots \tau_j = \nu > 0.$$

We have

$$|\psi^{(\tau)}(z)| \underset{\tau}{\leqslant} \int\limits_{\partial\mathbb{D}\setminus D(z,s_0)} \left|\log\left|\frac{f(t)}{P(t)}\right|\right| \frac{|dt|}{|t-z|^{\tau+1}} \leqslant \frac{1}{s_0^{\tau-1}} \int\limits_{\partial\mathbb{D}\setminus D(z,s_0)} \left|\log\left|\frac{f(t)}{P(t)}\right|\right| \frac{|dt|}{|t-z|^2} \leqslant$$

$$\underset{A,\alpha}{\leqslant} \frac{1}{s_0^{\tau}} \int\limits_{\partial\mathbb{D}} \left|\log\left|\frac{f(t)}{P(t)}\right|\right| \frac{1-|\tilde{z}|^2}{|t-z|^2} |dt|, \qquad \tilde{z} = \left(1 - \frac{s_0}{C_0^{1/\alpha} A^{1/\alpha}}\right) z_0 . \tag{1.5.6}$$

Lemma 1.3 and (1) imply

$$\int\limits_{\partial\mathbb{D}} \left|\log\left|\frac{f(t)}{P(t)}\right|\right| \frac{1-|\tilde{z}|^2}{|t-\tilde{z}|^2} |dt| \leqslant \int\limits_{\partial\mathbb{D}} \left|\log\left|\frac{f(t)}{M_f(\tilde{z})}\right|\right| \frac{1-|\tilde{z}|^2}{|t-\tilde{z}|^2} |dt| +$$

$$+ \int\limits_{\partial\mathbb{D}} \left|\log\left|\frac{P(t)}{M_f(\tilde{z})}\right|\right| \frac{1-|\tilde{z}|^2}{|t-\tilde{z}|^2} |dt| \underset{A,n}{\leqslant} 1, \tag{1.5.7}$$

since by the definition of $s(A; P, z)$ we have

$$M_P(\tilde{z}) \geqslant C_{\alpha 0} \max_{D(z,s)} |P| = C_{\alpha 0} A s^{\alpha},$$

and then our choice of A yields

$$M_f(\tilde{z}) \underset{A,n}{\asymp} M_P(\tilde{z}), \quad M_f(\tilde{z}) \geqslant (1-|\tilde{z}|)^d,$$

and we can obtain (1.5.7) in the same ways (1.5.5). Now from (1.5.6) and (1.5.7) we deduce

$$|\varphi^{(r)}(z)| \underset{A,d,r}{\preccurlyeq} s_0^{-r}$$

and hence

$$|\varphi_{s_0}^{(r)}(z)| \underset{A,d,r}{\preccurlyeq} s_0^{-r}, \quad r = 1, 2, \dots . \quad \bullet$$

1.6. The proof of Theorem 5: the part b).

We complete the proof of the part b) of Theorem 5 if we prove the estimate

$$|f^{(n+1)}(z)| \leqslant C(1-|z|)^{d-n-1}, \quad z \in \mathbb{D} \tag{1.6.1}$$

where C does not depend on z, $f = {}_e F$, $F \in C^d(\partial\mathbb{D})$. Without loss of generality we may assume $\|F\|_{C^d} = 1$.

It is sufficient to prove (1.6.1) only for $\rho = 1 - |z| < 0{,}1$. Let $z_0 = z/|z|$, P be the Taylor polynomial of F at z_0, $\deg P \leqslant n$. We choose the number A in the inequalities (1.6.2) - (1.6.5) below. Then we use different arguments in the cases $M_F(z) \leqslant A\rho^d$ and $M_F(z) > A\rho^d$. We take

$$A = 4^{n+1} A_0, \quad A_0 \geqslant 1 \quad (\text{for } s(A_0; P, z) \geqslant 4\rho \qquad) \tag{1.6.2}$$

$$C_d(\text{Lemma 1.11}) \leqslant A_0\, \sigma_{nn} \quad (\text{Lemma 1.9}) \qquad (\text{for Lemma 1.9}) \tag{1.6.3}$$

$$C_d(\text{Lemma 1.11}) \leqslant A_0\, \sigma_n^0 \quad (\text{Lemma 1.10}) \qquad (\text{for Lemma 1.10}) \tag{1.6.4}$$

$$A_0 \geqslant A(d) \ (\text{Lemma 1.13}) \qquad (\text{for Lemma 1.13}) \tag{1.6.5}$$

CASE 1. $M_F(z) \leqslant A\rho^d$.

The Cauchy formula gives

$$|f^{(n+1)}(z)| \underset{n}{\lesssim} \rho^{-n-1} \max_{\partial D(z,\rho/2)} |f(\zeta)|. \tag{1.6.6}$$

If $\zeta \in \partial D(z, \rho/2)$ then Lemma 1.2 implies

$$\log|f(\zeta)| = \frac{1}{2\pi}\int_{\partial\mathbb{D}} \log|F(t)| \frac{1-|\zeta|^2}{|t-\zeta|^2}|dt| \leqslant$$

$$\leqslant \frac{1}{2\pi}\int_{\partial\mathbb{D}} \log(|P(t)|+|F(t)-P(t)|)\frac{1-|\zeta|^2}{|t-\zeta|^2}|dt| \leqslant$$

$$\leqslant \frac{1}{2\pi}\int_{\partial\mathbb{D}} \log(|P(t)|+\tilde{C}_d|t-z_0|^d)\frac{1-|\zeta|^2}{|t-\zeta|^2}|dt| \leqslant$$

$$\leqslant \log[C_{d1}A(\max_{D(z,2\rho)}|P|+\rho^d)] \leqslant \log(C_{d2}A\rho^d)$$

and then with account of (1.6.6) we have

$$|f^{(n+1)}(z)| \leqslant C_{d3}A\rho^{d-n-1} \tag{1.6.7}$$

CASE 2. $M_F(z) > A\rho^d$. Let Ξ be the zero set of P .
Using the Dirichlet-principle, we can find a set of annuli $\mathcal{L}=\{L_j\}_{j=0}^{N+1}$ with the following properties:

1. $N \leqslant 2n-1$ (1.6.8$_1$)

2. $L_0 = D(z, X_0)$, $L_j = D(z, X_{j-1}, X_j)$, $j \leqslant N$, (1.6.8$_2$)

$L_{N+1} = \mathbb{C}\setminus D(z, X_N)$, $\frac{4\rho}{3} \leqslant X_0$, $X_j < X_{j+1} < \infty$;

3. If $L_j \cap \Xi \neq \emptyset$ then $\mathcal{D}(z, q_n' X_{j-1}, q_n'' X_j) \cap \Xi = \emptyset$

where $q_n' = 1 - \frac{1}{4(n+1)^2}$, $q_n'' = 1 + \frac{1}{4(n+1)^2}$ (1.6.8$_3$)

If $L_j \cap \Xi \neq \emptyset$ then $\mathcal{D}(z, x_{j-1}, q_n'' x_{j-1}) \cap \Xi = \emptyset$, $\mathcal{D}(z, q_n' x_j, x_j) \cap \Xi = \emptyset$. The L_0 and L_{N+1} have similar properties.

4. If $L_j \cap \Xi = \emptyset$, $1 \leq j \leq N$ then

$$\frac{4}{3} \leq \frac{X_j}{X_{j-1}} \leq \frac{5}{2} \tag{1.6.8$_4$}$$

if $L_0 \cap \Xi \neq \emptyset$ then

$$X_0 \leq 2\rho,$$

if $L_j \cap \Xi = \emptyset$, $1 \leq j \leq N$, then

$$\frac{4}{3} \leq \frac{X_j}{X_{j-1}}.$$

We define the number s_0 as follows: if $s = s(A_0; P, z) \geq 2$ then $s_0 = 2$

if $s < 2$ and there is a circle ∂L_j in $\mathcal{D}(z, \frac{s}{3}, s)$, then s_0 is the greatest radius of such circles;

if $s < 2$ and $\mathcal{D}(z, \frac{s}{3}, s)$ does not contain any ∂L, then $s_0 = \frac{s}{2}$.

In the case $s_0 = 2$ our arguments reduce to the arguments used in the case $s_0 < 2$. So we may assume $s_0 < 2$. Let $\Xi = \Xi \cap D(z, s_0)$. Then the set

$$\mathcal{L}_0 = \{L_j \cap D(z, s_0), \mathbb{C} \setminus D(z, s_0)\}_{j=0}^{N}.$$

has the properties 1 - 4 for the point z and the set Ξ_0. We number the rings in $\mathcal{L}_0$ according to their distances to z: $L_0, \ldots, L_K, L_{K+1}, K \leq N, L_j = D(z, X_{j-1}, X_j), 1 \leq j \leq k$, L_0 is the disc $D(z, X_0)$, $L_{K+1} = \mathbb{C} \setminus D(z, s_0)$. Let m_j be the number of the zeros P in L_j, $0 \leq j \leq K+1$. If $m_j > 0$ and $\zeta_{j1}, \ldots, \zeta_{jm_j}$ are the zeros of P then we put

$$P_j(\zeta) = \prod_{l=1}^{m_j} (\zeta - \zeta_{jl}), \quad 0 \leq j \leq K, \quad m_j > 0; \tag{1.6.8$_5$}$$

$$P_j(\zeta) \equiv 1, \qquad 0 \leqslant j \leqslant K, \quad m_j = 0;$$

$$Q_j =_e P_j, \qquad 0 \leqslant j \leqslant K; \tag{1.6.8$_6$}$$

$$Q =_e P; \tag{1.6.8$_7$}$$

$$P_{K+1} = P \Big/ \prod_{j=0}^{K} P_j; \tag{1.6.8$_8$}$$

$$Q_{K+1} =_e P_{K+1};$$

$$I_j = \partial\mathbb{D} \cap L_j, \quad 0 \leqslant j \leqslant K+1. \tag{1.6.8$_9$}$$

We put for $0 \leqslant j \leqslant K$

$$\varphi_j(\zeta) \overset{def}{=} Q_j(\zeta) \cdot \exp \frac{1}{2\pi} \int_{I_j} \log \left| \frac{F(t)}{P(t)} \right| \frac{t+\zeta}{t-\zeta} |dt|; \tag{1.6.8$_{10}$}$$

$$\varphi_{K+1}(\zeta) = Q_{K+1}(\zeta) \cdot \exp \frac{1}{2\pi} \int_{I_{K+1}} \log \left| \frac{F(t)}{P(t)} \right| \frac{t+\zeta}{t-\zeta} |dt|. \tag{1.6.8$_{11}$}$$

We now obtain our main identity:

$$\begin{aligned} f(\zeta) &= \exp \frac{1}{2\pi} \int_{\partial\mathbb{D}} \log |F(t)| \frac{t+\zeta}{t-\zeta} |dt| = \\ &= Q(\zeta) \cdot \exp \frac{1}{2\pi} \int_{\partial\mathbb{D}} \log \left| \frac{F(t)}{P(t)} \right| \frac{t+\zeta}{t-\zeta} |dt| = \\ &= \prod_{j=0}^{K+1} Q_j(\zeta) \cdot \prod_{j=0}^{K+1} \exp \frac{1}{2\pi} \int_{\partial\mathbb{D} \cap L_j} \log \left| \frac{F(t)}{P(t)} \right| \frac{t+\zeta}{t-\zeta} |dt| = \\ &= \prod_{j=0}^{K+1} \varphi_j(\zeta). \end{aligned} \tag{1.6.8}$$

Differentiating this identity we obtain

$$f^{(n+1)}(\zeta)=\sum_{(\nu_0,\dots,\nu_{K+1})}\frac{(n+1)!}{\nu_0!\dots\nu_{K+1}!}\prod_{j=0}^{K+1}\varphi_j^{(\nu_j)}(\zeta), \tag{1.6.9}$$

where $\nu_j\geqslant 0,\ \nu_0+\dots+\nu_{K+1}=n+1$. Let us put in (1.6.9) $\xi=z$ and estimate one term $t_{\nu_0,\dots,\nu_{K+1}}$. Since $\sum_{j=0}^{K+1}\nu_j=n+1>n\geqslant \deg P=\sum_{j=0}^{K+1}m_j$ we can find such j_0 that $\nu_{j_0}>m_{j_0}$; let j_0 be the least of such numbers. We now write down the estimates for $\varphi_j^{(\nu)}$ which we have already obtained earlier. These estimates hold in our situation because of the choice of the number A in (1.6.2) - (1.6.5) and, because $M_F(z)>A\rho^d$. In the signs $\lessdot$ below omit indexes for brevity

$$\left.\begin{aligned}&|\varphi_0(z)|\lessdot 1\\&|\varphi_0^{(\nu)}(z)|\lessdot \sigma_0X_0^{-\nu}+\sigma_{-1}X_{-1}^{-\nu}\end{aligned}\right\}\quad \text{if } m_0=0 \tag{1.6.10}$$

$$\left.\begin{aligned}&|\varphi_0^{(\nu)}(z)|\lessdot X_0^{m_0-\nu},\quad 0\leqslant\nu\leqslant m_0\\&|\varphi_0^{(\nu)}(z)|\lessdot \sigma_0X_0^{m_0-\nu},\quad \nu>m_0,\end{aligned}\right\}\quad \text{if } m_0>0, \tag{1.6.11}$$

where in (1.6.10) and below $X_{-1}=\rho,\quad \sigma_j=\dfrac{X_j^d}{\max\limits_{D(z,X_j)}|P|},\quad -1\leqslant j\leqslant K.$

We have next for $1\leqslant j\leqslant K$

$$\left.\begin{aligned}&|\varphi_j(z)|\lessdot 1\\&|\varphi_j^{(\nu)}(z)|\lessdot \sigma_{j-1}X_{j-1}^{-\nu}+\sigma_jX_j^{-\nu}\end{aligned}\right\}\quad \text{if} \tag{1.6.12}$$

$$\left.\begin{aligned} &|\varphi_j^{(\nu)}(z)| \preccurlyeq X_j^{m_j-\nu}, \quad 0 \leq \nu \leq m_j \\ &\varphi_j^{(\nu)}(z) \preccurlyeq \sigma_j X_j^{m_j-\nu}, \quad \nu > m_j \end{aligned}\right\} \quad \text{if } m_j > 0. \tag{1.6.13}$$

Finally, we can write down the estimates for φ_{K+1} :

$$|\varphi_{K+1}^{(\nu)}(z)| \preccurlyeq Q_{K+1} X_K^{-\nu}, \quad \nu \geq 0, \tag{1.6.14}$$

where $Q_{K+1} = |Q_{K+1}(z)|$ (we remind that $X_K = s_0$). We estimate the term $t_{\nu_0,\dots,\nu_{K+1}}$ in (1.6.9). There are three cases.

1. $j_0 = K+1$. Since j_0 is the least index for which $\nu_{j_0} > m_{j_0}$ it follows that $\nu_j \leq m_j$, $0 \leq j \leq K$ and hence using (1.6.10) - (1.6.13) we obtain

$$|\varphi_j^{(\nu_j)}(z)| \preccurlyeq X_j^{m_j-\nu_j}, \quad m_j > 0, \quad 0 \leq j \leq K. \tag{1.6.15}$$

Since $\nu_j = 0$ implies $m_j = 0$, $0 \leq j \leq K$, we conclude from (1.6.14) and (1.6.15) that

$$t_{\nu_0,\dots,\nu_{K+1}} \preccurlyeq \prod_{0\leq j\leq K} X_j^{m_j-\nu_j} \cdot Q_{K+1} X_K^{-\nu_{K+1}} \leq \prod_{0\leq j\leq K} X_K^{m_j-\nu_j} \cdot Q_{K+1} X_K^{-\nu_{K+1}} =$$

$$= Q_{K+1} \prod_{\substack{0\leq j\leq K \\ m_j>0}} X_K^{m_j} \cdot X_K^{-n-1} \asymp \max_{D(z,X_K)} |Q(\zeta)| \cdot X_K^{-n-1} \asymp$$

$$\asymp X_K^{d-n-1} \leq \rho^{d-n-1}. \tag{1.6.16}$$

Since according to Lemmas 1.1 and 1.4 $\rho \leq X_{j-1} < X_j \leq X_K$, $0 \leq j \leq K$, $m_j - \nu_j \geq 0$ and according to the property

$$\max_{D(z,X_j)} |Q(\zeta)| \asymp \max_{D(z,X_j)} |P(\zeta)|, \quad -1 \leq j \leq K,$$

which follows from Lemma 1.1 and the fact that $Q(\zeta) = {}_e P(\zeta)$).

2. $j_0 < K+1, \; m_{j_0} > 0$. For $0 \leq j \leq j_0 - 1$ if $j_0 > 0$ then $\nu_j \leq m_j, \; 0 \leq j \leq j_0 - 1$. We use the estimate (1.6.13) for the factor $\varphi_{j_0}^{(\nu_0)}$ while for all other factors we use the estimate (1.6.15). We obtain the inequalities

$$t_{\nu_0,\ldots,\nu_{K+1}} \leq \prod_{0 \leq j \leq j_0-1} X_j^{m_j-\nu_j} \cdot \sigma_{j_0} X_{j_0}^{m_{j_0}-\nu_{j_0}} \cdot \prod_{j_0 < j \leq K} X_{j-1}^{m_j-\nu_j} \cdot Q_{K+1} X_K^{-\nu_{K+1}} \leq$$

$$\leq \prod_{0 \leq j \leq j_0-1} X_{j_0}^{m_j-\nu_j} \cdot \sigma_{j_0} X_{j_0}^{m_{j_0}-\nu_{j_0}} \cdot \prod_{j_0 < j \leq K} X_j^{m_j} \cdot \prod_{j_0 < j \leq K} X_{j_0}^{-\nu_j} \cdot Q_{K+1} X_{j_0}^{-\nu_{K+1}} =$$

$$= \sigma_{j_0} Q_{K+1} \prod_{j_0 < j \leq K} X_j^{m_j} \cdot \prod_{0 \leq j \leq j_0} X_{j_0}^{m_j} \cdot X_0^{-n-1} \asymp \sigma_{j_0} \max_{D(z,X_{j_0})} |Q(\zeta)| \cdot X_{j_0}^{-n-1} \asymp$$

$$\asymp X_{j_0}^{d-n-1} \leq \rho^{d-n-1}. \tag{1.6.17}$$

3. $j_0 < K+1, \; m_{j_0} = 0$. We put $X_{-1} = \rho$. If $0 \leq j < j$ then $\nu_j \leq m_j$. We use the estimate (1.6.12) for the factor $\varphi_{j_0}^{(\nu_{j_0})}$ (if $j_0 = 0$ then we use (1.6.10)) and we use (1.6.15) for all other factors. Then we have

$$t_{\nu_0,\ldots,\nu_{K+1}} \leq \prod_{0 \leq j \leq j_0-1} X_j^{m_j-\nu_j} \cdot (\sigma_{j_0-1} X_{j_0-1}^{-\nu_{j_0}} + \sigma_j X_{j_0}^{-\nu_{j_0}}) \cdot \prod_{j_0 < j \leq K} X_{j-1}^{m_j-\nu_j} \cdot Q_{K+1} X_K^{-\nu_{K+1}} =$$

$$= \prod_{0 \leq j \leq j_0-1} X_j^{m_j-\nu_j} \cdot \sigma_{j_0-1} X_{j_0-1}^{-\nu_{j_0}} \cdot \prod_{j_0 < j \leq K} X_{j-1}^{m_j-\nu_j} \cdot Q_{K+1} X^{-\nu_{K+1}} +$$

$$+ \prod_{0 \leq j \leq j_0-1} X_j^{m_j-\nu_j} \cdot \sigma_{j_0} X_{j_0}^{-\nu_{j_0}} \cdot \prod_{j_0 < j \leq K} X_{j-1}^{m_j-\nu_j} \cdot Q_{K+1} X_K^{-\nu_{K+1}} \leq$$

$$\leq \prod_{0 \leq j \leq j_0-1} X_{j_0-1}^{m_j-\nu_j} \cdot \sigma_{j_0-1} X_{j_0-1}^{-\nu_{j_0}} \cdot \prod_{j_0 < j \leq K} X_j^{m_j} \cdot \prod_{j_0 < j \leq K} X_{j_0-1}^{-\nu_j} \cdot Q_{K+1} X_{j_0-1}^{-\nu_{K+1}} +$$

$$+\prod_{0\leq j\leq j_0-1} X_{j_0}^{m_j-\nu_j}\cdot \sigma_{j_0} X_{j_0}^{-\nu_{j_0}}\cdot \prod_{j_0<j\leq K} X_j^{m_j}\cdot \prod_{j_0<j\leq K} X_{j_0}^{-\nu_j}\cdot Q_{K+1} X_{j_0}^{-\nu_{K+1}} =$$

$$=\sigma_{j_0-1} Q_{K+1} \prod_{j_0<j\leq K} X_j^{m_j}\cdot \prod_{0\leq j\leq j_0-1} X_{j_0-1}^{m_j}\cdot X_{j_0-1}^{-n-1} \asymp$$

$$\asymp \sigma_{j_0-1} \max_{D(z,X_{j_0}-1)} |Q(\zeta)|\cdot X_{j_0-1}^{-n-1} + \sigma_{j_0} \max_{D(z,X_{j_0})} |Q(\zeta)|\cdot X_{j_0}^{-n-1} \leq$$

$$\leq X_{j_0-1}^{d-n-1} + X_{j_0}^{d-n-1} \leq \rho^{d-n-1}. \tag{1.6.18}$$

The relations (1.6.17) - (1.6.18) completes the proof of Theorem 5. ●

§ 2. The outer functions in H_n^p

THEOREM 6. a) Let $f\in H_n^p$, $n\geq 1$, $1<p<\infty$, $\psi(\zeta)=(f^{(n)})^*(\zeta)$, $\zeta\in\partial\mathbb{D}$.

Then if $M_f(z)\geq C_n\psi(z_0)\rho^n$ with constant C_n depending only on n then f satisfies (1(at z))

b) Suppose that some functions ψ, f,

$$\psi\in L^p(\partial\mathbb{D}) \quad \text{and} \quad f\in L_n^p(\partial\mathbb{D})$$

satisfy

$$M_f(z)\geq\psi(z_0)\rho^n \quad \Rightarrow \qquad \text{f satisfies (1) at } z.$$

Then $ef\in H_n^p$.

2. The preparation.

We need some wellknown facts.

LEMMA 2.1. Let

$$f \in L_n^p(\partial\mathbb{D}), \quad 1 \leqslant p \leqslant \infty, \qquad \psi = (f^{(n)})^*,$$

P be the Taylor polynomial of f at ξ, $\deg P \leqslant n-1$, $\xi \in \partial\mathbb{D}$.

Then

$$|f(\zeta) - P(\zeta)| \leqslant C_{n0}|\zeta - \xi|^n \psi(\xi), \quad \zeta \in \partial\mathbb{D}. \quad \bullet \tag{2.0.1}$$

LEMMA 2.2. Let

$$f \in C^n[a,b], \quad x_\nu = x_0 + \nu h \in [a,b],$$

$$\nu = 1,\dots,n, \quad |f^{(n)}(x)| \leqslant A, \quad x \in [a,b].$$

Then

$$\Big|\sum_{\nu=0}^{n-j} (-1)^\nu C_{n-j}^\nu f^{(j)}(x_\nu)\Big| \leqslant C_{n1} h^{n-j} A, \quad 0 \leqslant j \leqslant n. \quad \bullet \tag{2.0.2}$$

2.1. Part a) of the Theorem 6.

We choose the number C_n in (2.1.7) later. Suppose that

$$M_f(z) \geqslant C_n \psi(z_0)\rho^n, \quad \psi(\zeta) = (f^{(n)})^*(\zeta).$$

Let $M = M_f(z)$, $\psi = \psi(z_0)$, P be the Taylor polynomial of f at z_0, $\deg P \leqslant n-1$. We note that for $2^n C_{n0} \leqslant \frac{1}{4} C_n$ we have

$$\max_{0 \leqslant j \leqslant n-1} |f^{(j)}(z_0)|\rho^j \geqslant C_3 M, \quad C_3 = \frac{1}{4}e^{-2}, \tag{2.1.1}$$

since otherwise for $z_* \in \gamma(z)$ satisfying $|f(z_*)| = M$ in view of (2.0.1) and $|z_* - z_0| \leqslant 2\rho$ we would have

$$M = |f(z_*)| \leqslant |P(z_*)| + 2^n C_{n0}\psi\rho^n \leqslant \sum_{j=0}^{n-1} \frac{|f^{(j)}(z_0)|}{j!}(2\rho)^j +$$

$$+2^n C_{n0}\varphi\rho^n < \sum_{j=0}^{n-1} C_3 M\rho^{-j}\frac{2^j\rho^j}{j!} + 2^n C_{n0}\varphi\rho^n < C_3 e^2 M + \frac{1}{4}M \leqslant \frac{1}{2}M.$$

Let j_0, $0 \leqslant j_0 \leqslant n-1$ satisfies $|f^{(j_0)}(z_0)|\rho^{j_0} \geqslant C_3 M$. Let

$$z_\nu = (1-\nu\rho)z_0, \quad \nu = 1,\dots,n, \quad Q = {}_eP.$$

If $C_n \geqslant 2^{n+2} C_{n0}$ then $M_P(z) \asymp M_f(z)$ which as in 1.1 implies

$$|Q(\zeta)| \underset{n}{\asymp} M, \quad \zeta \in \Omega(z), \tag{2.1.2}$$

where

$$\Omega(z) = \{\zeta \in \mathbb{D} : \frac{\rho}{2} \leqslant 1-|\zeta| \leqslant (n+1)\rho, \ |\arg \frac{\zeta}{z_0}| \leqslant \pi\rho\}.$$

As in § 1 using Lemma 1.1 we obtain

$$\int\limits_{|f|\geqslant|P|} \log\left|\frac{f(t)}{P(t)}\right| \frac{1-|\zeta|^2}{|t-\zeta|^2}|dt| \leqslant C_{n4}, \quad \zeta \in \Omega(z). \tag{2.1.3}$$

We put now

$$B = \int\limits_{|f|<|P|} \log\left|\frac{P(t)}{f(t)}\right| \frac{1-|z|^2}{|t-z|^2}|dt|,$$

and using (1.1.2) - (1.1.3) we obtain from (2.12) - (2.13):

$$|f(\zeta)| \leqslant |Q(\zeta)|\cdot \exp\frac{1}{2\pi}\int\limits_{|f|<|P|} \log\left|\frac{f(t)}{P(t)}\right| \frac{1-|\zeta|^2}{|t-\zeta|^2}|dt| \cdot \exp\frac{1}{2\pi}\int\limits_{|f|\geqslant|P|} \log\left|\frac{f(t)}{P(t)}\right| \frac{1-|\zeta|^2}{|t-\zeta|^2}|dt| \leqslant$$

$$\leqslant C_{n5} M e^{-C_{6n}B}, \quad \zeta \in \Omega(z) \tag{2.1.4}$$

which together with the Cauchy-inequalities

$$|f^{(j_0)}(z_\nu)| \leq \frac{2^{j_0} j_0!}{\rho^{j_0}} \max_{|\zeta - z_\nu| = \frac{\rho}{2}} |f(\zeta)|, \quad \nu = 1, \dots, n$$

implies

$$|f^{(j_0)}(z_\nu)| \leq C_{n7} M \rho^{-j_0} e^{-C_{n6} B}, \quad \nu = 1, \dots, n. \tag{2.1.5}$$

We now deduce from (2.1.1), (2.0.2) and (2.1.5)

$$c_3 \rho^{-j_0} M \leq |f^{(j_0)}(z_0)| \leq \left|\sum_{\nu=0}^{n-j_0} (-1)^n C_{n-j_0}^{\nu} f^{(j_0)}(z_\nu)\right| +$$

$$+ \left|\sum_{\nu=1}^{n-j_0} (-1)^{\nu} C_{n-j_0}^{\nu} f^{(j_0)}(z_\nu)\right| \leq C_{n1} \sup_{0<r<1} |f^{(n)}(r z_0)| \cdot \rho^{n-j_0} +$$

$$+ 2^n C_{n7} M \rho^{-j_0} e^{-C_{n6} B}. \tag{2.1.6}$$

Since $\sup_{0<r<1} |f^{(n)}(r z_0)| \leq c\,\psi$, it follows from (2.1.6) that

$$M \leq C_{n8} \psi \rho^n + C_{n9} e^{-C_{n6} B} M,$$

which yields $B \underset{n}{\asymp} 1$ and finishes the proof in case $C_n \geq 4 C_{n3}$. Now we have only to choose C_n property. We put

$$C_n = 4(C_{n8} + 2^n C_{n0}).$$

The part a) of Theorem 6 is proved. ●

2.2. The first crucial estimate.

The proof of b) of Theorem 6 follows the proof of Theorem 5 with some modifications.

LEMMA 2.1 Let

$$f\in L_n^p(\partial\mathbb{D}),\quad 1\leqslant p<\infty,\quad P(\cdot)=P_{n-1}(f;z_0;\cdot),$$

$$\varphi=(f^{(n)})^*(z_0),$$

L be the annulus $D(z,X,X)$. Suppose that P has m, $0<m\leqslant n-1$ zeros $\xi_1,\dots,\xi_m$ in L and that the annuli $D(z,x,q_n''x)$ and $D(z,q_n'X,X)$ do not containt the zeros of P . Suppose further that

$$\frac{4}{3}\leqslant\frac{X}{x}\leqslant\frac{5}{2},\quad I=L\cap\partial\mathbb{D},\quad \gamma=\partial\mathbb{D}\cap D(z,X)_m$$

and put

$$\sigma=\frac{\varphi X^n}{\max\limits_{D(z,X)}|P|},\quad P_0(\zeta)=\prod_1^m(\zeta-\zeta_\nu),\quad Q={}_eP_0,$$

$$\varphi(\zeta)\overset{def}{=\!=}Q(\zeta)\exp\frac{1}{2\pi}\int\limits_I\log\left|\frac{f(t)}{P(t)}\right|\frac{t+\zeta}{t-\zeta}|dt|.$$

Then there exists $\sigma_n>0$ such that if $\sigma\leqslant\sigma_n$ then

$$|\varphi^{(\nu)}(z)|\underset{\nu,n}{\lessdot}X^{m-\nu},\quad \nu\geqslant 0,\tag{2.2.1}$$

$$|\varphi^{(\nu)}(z)|\underset{\nu,n}{\lessdot}\sigma X^{m-\nu},\quad \nu\geqslant m+1.\tag{2.2.2}$$

PROOF. We use Lemma 1.9 from § 1. Let $\xi_1,\dots,\xi_m$ be zeros of P in the annulus L . We put $f-P=V$ and

$$R(\zeta)=\frac{P(\zeta)}{P_0(\zeta)}.$$

Then we have

$$\frac{f}{P}=1+\frac{V}{P_0R}=1+\frac{V/R}{P_0}.$$

The Taylor formula yields

$$\left|\left(\frac{V}{R}\right)^{(\nu)}(z)\right| \underset{\nu,n}{\lesssim} \sigma X^{m-\nu}, \qquad 0 \leqslant \nu \leqslant n-1.$$

Now Lemma 2.1 follows from Lemma 1.9. ●

2.3. The second crucial estimate.

We recall the notation

$$H^*\varphi(\zeta) \stackrel{def}{=} \sup_{r>0} \left| \int\limits_{\partial\mathbb{D}\setminus D(\zeta,r)} \frac{\varphi(t)}{t-\zeta} |dt| \right|.$$

LEMMA 2.2. Let

$$f \in L^p_n(\partial\mathbb{D}), \quad 1<p<\infty, \quad n \geqslant 1, \quad P = P_{n-1}(f; z; \cdot).$$

Suppose that P has no zeros in the ring $L_0 = D(z, q'_n x, q''_n X)$, where $x \geqslant 4/3\rho$, $X/x \geqslant 4/3$, $L = D(z, x, X)$,

$$I = \partial\mathbb{D} \cap L, \quad \varphi = (f^{(n)})^*(z_0), \quad g = (H^* f^{(n)})(z_0).$$

Suppose that P has $m \geqslant 0$ zeros in the disc $D(z,x)$ and let

$$\Psi(\zeta) \stackrel{def}{=} \frac{1}{2\pi} \int\limits_I \log\left|\frac{f(t)}{P(t)}\right| \frac{t+\zeta}{t-\zeta} |dt|,$$

$$\sigma^*_X = \frac{(\varphi+g)X^n}{\max\limits_{D(z,X)} |P|}, \quad \sigma_y = \frac{\varphi y^n}{\max\limits_{D(z,y)} |P|}, \quad y>0.$$

Then there exists $\sigma_{n0} > 0$ such that if $\sigma_X \leqslant \sigma_{n0}$ then

$$|\mathrm{Re}\,\Psi(z)| \lesssim 1 \tag{2.3.1'}$$

$$|\Psi^{(\nu)}(z)| \underset{\nu,n}{\lesssim} \sigma_x x^{-\nu} + \sigma_X X^{-\nu}, \quad \nu \geqslant 1, \quad \nu \neq n-m, \tag{2.3.1}$$

$$|\Psi^{(n-m)}(z)| \underset{n}{\lesssim} \sigma_X X^{m-n} \log\frac{X}{x}, \tag{2.3.2}$$

$$|\Psi^{(\nu)}(z)| \underset{n,\nu}{\preccurlyeq} x^{-\nu}, \quad \nu \geqslant 1, \tag{2.3.3}$$

$$|\Psi^{(n-m)}(z)| \underset{n}{\preccurlyeq} \sigma_X^* X^{m-n}. \tag{2.3.4}$$

PROOF. Let $\xi_1, \dots, \xi_m$ be zeros of P in the disc $\mathfrak{D}(z,x)$

$$r(\zeta) = \prod_1^m (\zeta - \zeta_\nu), \quad R(\zeta) = P(\zeta)/r(\zeta), \qquad R = |R(z)|.$$

The Taylor formula and the properties of the zeros of P give

$$\left|\frac{f(\zeta)}{P(\zeta)} - 1\right| \underset{n}{\preccurlyeq} \frac{\varphi \cdot |\zeta - z_0|^n}{R \cdot |\zeta - z_0|^m}, \quad \zeta \in L. \tag{2.3.5}$$

Hence for $\sigma_X \leqslant \sigma_{n0}$ we have

$$\left|\frac{f(\zeta)}{P(\zeta)} - 1\right| \leqslant \frac{1}{2}, \qquad \zeta \in I$$

and then for such a σ_{n0} we get from (2.3.5)

$$|\Psi^{(\nu)}(z)| \underset{n}{\preccurlyeq} \int_I \left|\log\left|\frac{f(t)}{P(t)}\right|\right| \frac{|dt|}{|t-z|^{\nu+1}} \preccurlyeq$$

$$\preccurlyeq \int_I \frac{\varphi}{R} \frac{|dt|}{|t-z|^{\nu+m+1-n}} \preccurlyeq \begin{cases} \dfrac{\varphi}{R}\left(x^{n-m-\nu} + X^{n-m-\nu}\right), & \nu \neq n-m, \\[2ex] \dfrac{\varphi}{R} \log\dfrac{X}{x}, & \nu = n-m. \end{cases} \tag{2.3.6}$$

Let us state now (2.3.3) and (2.3.4). We may assume $\sigma_{n0} \leqslant 1$. We have

$$\frac{\varphi}{R} y^{n-m-\nu} = \frac{\varphi y^n}{R y^m} y^{-\nu} \leqslant \sigma_y y^{-\nu} \leqslant y^{-\nu}, \qquad y = x \text{ or } X \tag{2.3.7}$$

$$\frac{\varphi}{R}\log\frac{X}{x}=\frac{\varphi X^{n}}{RX^{m}}X^{m-n}\log\frac{X}{x}\leqslant X^{m-n}\log\frac{X}{x}\leqslant x^{m-n}; \qquad (2.3.7')$$

Now (2.3.7') imply (2.3.3). We must to check only (2.3.4). We write

$$\varphi^{(n-m)}(z)=c\int_I\frac{f(t)-P(t)}{P(t)}\frac{|dt|}{(t-z)^{n-m+1}}+$$

$$+O\left(\int_I\left|\frac{f(t)-P(t)}{P(t)}\right|^2\frac{|dt|}{|t-z|^{n-m+1}}\right)=$$

$$=c\int_I\frac{f(t)-P(t)}{P(t)}\frac{|dt|}{(t-z)^{n-m+1}}+O\left(\int_I\frac{\varphi^2}{R^2}|t-z_0|^{n-m-1}|dt|\right)=$$

$$=c\int_I\frac{f(t)-P(t)}{P(t)}\frac{|dt|}{(t-z)^{n-m+1}}+O\left(\frac{\varphi}{R}\right), \qquad (2.3.8)$$

because $\frac{\varphi}{R}X^{n-m}\leqslant 1$. Let

$$v=f-P,\quad P_\nu(t)=\frac{t-z}{t-\zeta_\nu}P(t).$$

Then

$$\int_I\frac{v(t)}{P(t)}\frac{|dt|}{(t-z)^{n-m+1}}=\int_I\frac{v(t)}{P(t)}\frac{t-\zeta_\nu}{t-z}\frac{|dt|}{(t-z)^{n-m+1}}+$$

$$+\int_I\frac{v(t)}{P(t)}\frac{\zeta_\nu-z}{(t-z)^{n-m+2}}|dt|=\int_I\frac{v(t)}{P_\nu(t)}\frac{|dt|}{(t-z)^{n-m+1}}+O\left(\frac{\varphi}{R}\right). \qquad (2.3.9)$$

Applying (2.3.9) for $\nu=1,\dots,m$, we find

$$\int_I\frac{v(t)}{P(t)}\frac{|dt|}{(t-z)^{n-m+1}}=\int_I\frac{v(t)}{P_0(t)}\frac{|dt|}{(t-z)^{n-m+1}}+O\left(\frac{\varphi}{R}\right), \qquad (2.3.10)$$

where $P_0(t)=(t-z)^m R(t)$. Let $\xi_1^0,\dots,\xi_{n-m+1}^0$ be the zeros of $R(t)$ (they be in $\mathbb{C}\setminus D(z,q''_n X)$). Let

$$P_{0\nu}(t)=\frac{z-\zeta_\nu^0}{t-\zeta_\nu^0}P_0(t).$$

Then

$$\int_I \frac{v(t)}{P_0(t)}\frac{|dt|}{(t-z)^{n-m+1}}=\int_I \frac{v(t)}{P_{0\nu}(t)}\frac{|dt|}{(t-z)^{n-m+1}}+$$

$$+\int_I \frac{v(t)}{P_0(t)}\frac{|dt|}{(\zeta_\nu^0-z)(t-z)^{n-m}}=\int_I \frac{v(t)}{P_{0\nu}(t)}\frac{|dt|}{(t-z)^{n-m+1}}+O\left(\frac{\psi}{R}\right) \qquad (2.3.11)$$

because $\zeta_\nu^0\in\mathbb{C}\setminus D(z,q''_n X)$. Applying (2.3.11) for $\nu=1,\dots,n-m-1$ and taking into account (2.3.8) and (2.3.9) we get

$$\int_I \log\left|\frac{f(t)}{P(t)}\right|\frac{|dt|}{(t-z)^{n-m-1}}=\int_I \frac{v(t)}{P_{00}(t)}\frac{|dt|}{(t-z)^{n-m-1}}+O\left(\frac{\psi}{R}\right), \qquad (2.3.12)$$

where $P_{00}(t)=Re^{i\theta}(t-z)^m$. Integrating in (2.3.12) by parts we obtain

$$|\Psi^{(n-m)}(z)|\underset{n}{\lesssim}\frac{g+\psi}{R}\asymp\sigma_X^* X^{m-n}. \qquad \bullet$$

LEMMA 2.2'. The estimates (2.3.1') - (2.3.4) hold if L is the disc $D(z,X)$, $X\geqslant 2\rho$, $I=L\cap\partial\mathbb{D}$ the polynomial P has no zeros in $D(z,q''_n X)$ and in the formula (2.3.1') - (2.3.4) $x=\rho$.

The proof is identical to that of Lemma 2.2.

COROLLARY. In the assumptions of Lemma 2.2 or Lemma 2.2' we have

$$|\Psi^{(\nu)}(z)|\lesssim\sigma_x^* x^{-\nu}+\sigma_X^* X^{-\nu},\ \nu\geqslant 1. \qquad \bullet$$

LEMMA 2.3. In the assumptions of Lemma 2.2 or 2.2' for

$$\Phi(\zeta)=\exp\Psi(\zeta).$$

for $\sigma_X \leqslant \sigma_{n0}$ we have

$$|\Phi(z)| \leqslant 1,$$

$$|\Phi^{(\nu)}(z)| \underset{\nu,n}{\leqslant} \sigma_x^* x^{-\nu} + \sigma_X^* X^{-\nu}, \quad \nu \geqslant 1.$$

PROOF follows from (2.3.3), (2.3.4) similarly to (1.4.4) and (1.4.5). ●

LEMMA 2.4. Let $f \in L_n^p(\partial\mathbb{D})$, $1 < p < \infty$ and suppose that f satisfies the assumptions of Theorem 6 with a function ψ,

$$\psi \in L^p(\partial\mathbb{D}), \quad \omega = \psi + (f^{(n)})^* + H^* f^{(n)}.$$

Then there exists a number A_{n0} depending only on n such that it

$$M_f(z) \geqslant A_{n0}\,\omega(z_0)\rho^n$$

and

$$\varphi(\zeta) \overset{\text{def}}{=} \exp \frac{1}{2\pi} \int_{\partial\mathbb{D}\setminus D(z,2\rho)} \log\left|\frac{f(t)}{P(t)}\right| \frac{t+\zeta}{t-\zeta}\,|dt|, \quad P = P_{n-1}(f; z_0; \cdot)$$

then

$$|\varphi^{(\nu)}(z)| \underset{\nu,n}{\leqslant} \rho^{-n}, \quad \nu \geqslant 0.$$

PROOF. We notice that

$$|f(t)| - P_{n-1}(f; z_0; t) \underset{n}{\leqslant} (f^{(n)})^*(z_0)\cdot|t-z_0|^n.$$

Then if we have the inequality

$$M_f(z) \geqslant A_{n0}\,\omega(z_0)\rho^n$$

for a proper A_{n0}, then

$$M_f(z) \asymp M_P(z), \qquad P = P_{n-1}(f; z; \cdot)$$

and we finish the proof literally as in 1.13. ●

2.4. Proof of the part b) of Theorem 6.

Let

$$\omega = \psi + (f^{(n)})^* + H^* f^{(n)}.$$

We prove the estimate

$$|({}_e f)^{(n)}(z)| \underset{n,f,\psi}{\preccurlyeq} \omega(z), \quad z \in \mathbb{D}, \tag{2.4.1}$$

CASE 1. $M_f(z) \leqslant A_n \, \omega(z_0) \rho^n$. A_n depends only on n .Then

$$|({}_e f)^{(n)}(z)| \underset{n}{\preccurlyeq} \rho^{-n} \max_{\partial D(z, \frac{\rho}{2})} |{}_e f(\zeta)|. \tag{2.4.2}$$

Let $P = P_{n-1}(f; z_0; \cdot)$. The Taylor formula gives

$$M_P(z) \preccurlyeq M_f(z) + \psi(z_0) \cdot \rho^n \preccurlyeq \omega(z_0) \rho^n, \tag{2.4.3}$$

and then for $\zeta \in \partial D(z, \frac{\rho}{2})$ by means of (2.4.3) and Lemma 1.2 we have

$$|{}_e f(\zeta)| \preccurlyeq \exp \frac{1}{2\pi} \int_{\partial \mathbb{D}} \log[|P(t)| + |t - z_0|^n \omega(z_0)] \cdot \frac{1 - |\zeta|^2}{|t - \zeta|^2} |dt| \preccurlyeq \rho^n \omega(z_0) \tag{2.4.4}$$

where the sign $\preccurlyeq$ in (2.4.4) depends only on n and A_n . Together with (2.4.2) this implies (2.4.1). We specify A_n by

$$A_n = 4^n \tilde{A}_n C_n,$$

C_n being the constant from the inequality

$$\max_{D(z,4\rho)} |P| \geq C_n \cdot 4^n \max_{D(z,2\rho)\cap\partial\mathbb{D}} |P|; \tag{2.4.5}$$

where $\tilde{A}_n$ is defined below in the definition of the number s.

CASE 2. $M_f(z) > A_n \omega(z_0)\rho^n$. Let

$$P = P_{n-1}(f; z_0; \cdot), \quad \omega = \omega(z_0).$$

We put s as follows:

$$s = s(\tilde{A}_n) = \max \{ \delta: \quad \text{for } 0 \leq \tau \leq \delta \quad \max_{D(z,\tau)} |P(t)| \geq \tilde{A}_n \omega \tau^n \}.$$

As in § 1 the case $s \geq 2$ is simpler. Therefore we assume $s < 2$. We take the number $\tilde{A}_n$, $\tilde{A}_n \geq 1$ depending only on n in such a way that for any X, $\rho \leq X \leq s$ the estimates (2.2.1) - (2.2.2) and (2.3.1) - (2.3.4) hold. It is clear that this is possible. Because $A_n = 4^n C_n \tilde{A}_n$ we have $s \geq 4\rho$. Let Ξ be the zero-set of the polynomial P. We choose the set of annuli $\{L_j\}_{j=0}^{N+1}$, $N \leq 2n-3$ in accordance with Ξ, ρ and s similarly to $(1.6.8_1)$ - $(1.6.8_4)$ of § 1,

$$L_j = D(z, X_{j-1}, X_j), \quad 1 \leq j \leq N, \quad L_0 = D(z, X_0), \quad X_N = s.$$

and define the functions φ_j, $0 \leq j \leq N+1$ as in (1.6.8) $-(1.6.8_{11})$ using the polynomial P, the function f and the set $\{L_j\}_{j=0}^{N+1}$. Then we write

$$ef(\zeta) = \prod_{j=0}^{N+1} \varphi_j(\zeta), \tag{2.4.6}$$

$$(ef)^{(n)}(z) = \sum_{\nu_0+\ldots+\nu_{N+1}=n} \frac{n!}{\nu_0!\ldots\nu_{N+1}!} \prod_{j=0}^{N+1} \varphi_j^{(\nu_j)}(z) \stackrel{def}{=} \sum_{\nu_0+\ldots+\nu_{N+1}=n} \frac{n!}{\nu_0!\ldots\nu_{N+1}!} t_{\nu_0,\ldots,\nu_{N+1}}. \tag{2.4.7}$$

We recall the estimates for the functions φ_j and for their derivatives. Let $Q_{N+1}(t)$ be the polynomial corresponding to L_{N+1} ,

$$Q_{N+1} \overset{def}{=} |Q_{N+1}(z)|, \quad X_{-1} \overset{def}{=} \rho, \quad \sigma_j^* = \frac{\omega X_j^n}{\max\limits_{D(z,X_j)} |P|}, \quad -1 \leq j \leq N.$$

m_j be the number of zeros of z in L_j, $0 \leq j \leq N+1$. Then we have:

if $m_j > 0, \quad 0 \leq j \leq N$ then

$$|\varphi_j^{(\nu)}(z)| \leq \begin{cases} X_j^{m_j-\nu}, & \nu \geq 0, \\ \sigma_j^* X_j^{m_j-\nu}, & \nu \geq m_j+1; \end{cases} \tag{2.4.8}$$

if $m_j = 0, \quad 0 \leq j \leq N$ then

$$|\varphi_j^{(\nu)}(z)| \leq \begin{cases} X_{j-1}^{-\nu}, & \nu \geq 0, \\ \sigma_{j-1}^* X_{j-1}^{-\nu} + \sigma_j^* X_j^{-\nu}, & \nu > 0; \end{cases} \tag{2.4.9}$$

Finally,

$$|\varphi_{N+1}^{(\nu)}(z)| \leq Q_{N+1} X_N^{-\nu}, \quad \nu \geq 0. \tag{2.4.10}$$

Let us estimate the term $t_{\nu_0,\dots,\nu_{N+1}}$, where

$$\sum_{j=0}^{N+1} \nu_j = n > n-1 \geq \deg P = \sum_{j=0}^{N+1} m_j .$$

There exists an index j, $0 \leq j \leq N+1$ such that $\nu_j > m_j$. Let j_0 be the least such index. We consider three cases.

1. $j_0 = N+1$. We have $\nu_j \leq m_j$, $0 \leq j \leq N$. Then (2.4.8)-(2.4.10) imply

$$t_{\nu_0,\dots,\nu_{N+1}} \leq \prod_{0 \leq j \leq N} X_j^{m_j-\nu_j} \cdot Q_{N+1} X_N^{-\nu_{N+1}} \leq \prod_{0 \leq j \leq N} X_N^{m_j-\nu_j} \cdot Q_{N+1} X_N^{-\nu_{N+1}} =$$

$$= Q_{N+1} \prod_{0\le j\le N} X_n^{m_j} \cdot X_N^{-n} \asymp \max_{D(z,X_n)} |Q(t)| \cdot X_N^{-n} \preccurlyeq \omega . \qquad (2.4.11)$$

Because $\frac{s}{3} \leqslant s_0 \leqslant s$ we have

$$\max_{D(z,s_0)} |Q(t)| \asymp \max_{D(z,s_0)} |P| \underset{n}{\asymp} \tilde{A}_n \omega s_0^n = \tilde{A}_n \omega X_N^n .$$

2. $j_0 \leqslant N,\ m_{j_0} > 0$. Using (2.4.8) - (2.4.10) once more we get

$$t_{\nu_0,\dots,\nu_{N+1}} \leqslant \prod_{0\le j<j_0} X_j^{m_j-\nu_j} \cdot \sigma_{j_0}^{*} X_{j_0}^{m_{j_0}-\nu_{j_0}} \cdot \prod_{j_0<j\le N} X_{j-1}^{m_j-\nu_j} \cdot Q_{N+1} X_N^{-\nu_{N+1}} \leqslant$$

$$\leqslant \prod_{0\le j<j_0} X_{j_0}^{m_j-\nu_j} \cdot \sigma_{j_0}^{*} X_{j_0}^{m_{j_0}-\nu_{j_0}} \cdot \prod_{j_0<j\le N} X_j^{m_j} \cdot \prod_{j_0<j\le N} X_{j_0}^{-\nu_j} \cdot Q_{N+1} X_{j_0}^{-\nu_{N+1}} \asymp$$

$$\asymp \sigma_{j_0}^{*} \max_{D(z,X_{j_0})} |Q(t)| \cdot X_{j_0}^{-n} \asymp \omega . \qquad (2.4.12)$$

3. $j_0 \leqslant N,\ m_{j_0} = 0$. We write by means of (2.4.9):

$$t_{\nu_0,\dots,\nu_{N+1}} \leqslant \prod_{0\le j<j_0} X_j^{m_j-\nu_j} (\sigma_{j_0-1}^{*} X_{j_0-1}^{-\nu_j} + \sigma_{j_0}^{*} X_{j_0}^{-\nu_{j_0}}) \prod_{j_0\le j\le N} X_{j-1}^{m_j-\nu_j} \cdot Q_{N+1} X_N^{-\nu_{N+1}} \leqslant$$

$$\leqslant \sum_{\delta=0}^{1} \prod_{0\le j} X_{j_0-\delta}^{m_j-\nu_j} \cdot \sigma_{j_0-\delta}^{*} X_{j_0-\delta}^{-\nu_{j_0}} \cdot \prod_{j_0+1\le j\le N} X_j^{m_j} \cdot \prod_{j_0+1\le j\le N} X_{j_0-\delta}^{-\nu_j} \cdot Q_{N+1} X_{j_0-\delta}^{-\nu_{N+1}} \asymp$$

$$\asymp \sum_{\delta=0}^{1} \sigma_{j_0-\delta}^{*} X_{j_0-\delta}^{-n} \cdot \max_{D(z,X_{j_0-\delta})} |Q(t)| \preccurlyeq \omega . \qquad (2.4.13)$$

Now (2.4.11) - (2.4.13) finish the proof of the theorem 6. ●

§ 3. The class $\Lambda^{n-1}Z$.

THEOREM 7. a) Let $f\in\Lambda^{n-1}Z$, $n\geqslant 1$, $M_f(z)\geqslant\rho^n$. Then (1) holds for f at z;
b) Let $f\in C^{n-1}Z(\partial\mathbb{D})$ and let (1) holds for all $z\in\mathbb{D}$ such that $M_f(z)\geqslant\rho^n$. Then $_ef\in\Lambda^{n-1}Z$. We begin the proof with some technical facts.

3.0. Some necessary properties of the class $C^{n-1}Z$.

LEMMA 3.1. Let $I\subset\mathbb{R}$, $F\in C^{n-1}Z(I)$, $n\geqslant 2$,

$$|\Delta^2F^{(n-1)}(x,h)|\overset{def}{=}|F^{(n-1)}(x)-2F^{(n-1)}(x+h)+F^{(n-1)}(x+2h)|\leqslant\lambda h$$

for any $x,\ x+h,\ x+2h\in I$. Suppose that $x_0\in I$, $F(x_0)=0$, $f(x)=F(x)/(x-x_0)$. Then $f\in C^{n-2}Z(I)$ and

$$|\Delta^2f^{(n-2)}(x,h)|\leqslant c_n\lambda h. \tag{3.0.1}$$

PROOF. We may assume without loss of generality that $I=[0,1]$ and that F is defined on $[-1,2]$ in such a way that

$$|\Delta^2F^{(n-1)}(x,h)|\leqslant C_{n0}\lambda h,\quad x,x+h,x+2h\in[-1,2]$$

–see [32], Ch.3. Using classical theorems about approximation on a segment [32], Ch. 5 we may find for any $N\geqslant n$ a polynomial Q_N, $\deg Q_N\leqslant N$, with the following properties:

$$Q_n(x_0)=0; \tag{3.0.2}$$

$$|F^{(\nu)}(x)-Q_N^{(\nu)}(x)|\leqslant C_{n1}\lambda N^{\nu-n};\quad \nu=0,1,\dots,n-1,\quad x\in[0,1]; \tag{3.0.3}$$

$$|Q_N^{(n+1)}(x)|\leqslant C_{n1}\lambda N,\quad x\in[0,1]. \tag{3.0.4}$$

Suppose now that $0<h<1$, $x,x+h,\ x+2h\in I$. We put $N=\left[\frac{n}{h}\right]+1$, take the polynomial Q_N with the properties (3.0.2) –

(3.0.4) and put $\varphi_N = F - Q_N$, $P = P_{n-1}(\varphi_N, x_0, \cdot)$. Then

$$f(x) = \frac{\varphi_N(x)}{x-x_0} + \frac{Q_N(x)}{x-x_0},$$

$$\Delta^2 f^{(n-2)}(x,h) = \Delta^2\left(\frac{\varphi_N(x)}{x-x_0}\right)^{(n-2)}(x,h) + \Delta^2\left(\frac{Q_N(x)}{x-x_0}\right)^{(n-2)}(x,h) \overset{\text{def}}{=} T_1 + T_2. \qquad (3.0.5)$$

We estimate the terms T_1 and T_2 in different ways. Since $F(x_0) = Q_N(x_0) = 0$ we find $P(x_0) = 0$. Hence $P(x)/(x-x_0)$ is a polynomial of the degree $\leqslant n-2$. Hence

$$\Delta^2 (P(x)/(x-x_0))^{(n-2)}(x,h) \equiv 0$$

and therefore

$$T_1 = \Delta^2\left(\frac{\varphi_N(x) - P(x)}{x-x_0}\right)^{(n-2)}(x,h) =$$

$$= \Delta^2\Big[\sum_{\nu=0}^{n-2} (-1)^\nu C_{n-2}^\nu \frac{1}{(x-x_0)^{\nu+1}} \int_{x_0}^{x} (x-t)^\nu (\varphi_N^{(n-1)}(t) - \varphi_N^{(n-1)}(x_0))\,dt\Big](x,h). \qquad (3.0.6)$$

Let $y = x,\ x+h,\ x+2h$. Then (3.0.3) implies

$$\frac{1}{|y-x_0|^{\nu+1}}\left|\int_{x_0}^{y} |y-t|^\nu (|\varphi_N^{(n-1)}(t)| + |\varphi_N^{(n-1)}(x_0)|)\,dt\right| \underset{n}{\leqslant} \frac{\lambda}{N} \asymp \lambda h,$$

i.e. $|T_1| \underset{n}{\leqslant} \lambda h$. Let us estimate T_2. Let

$$q = P_n(Q_N; x_0, \cdot).$$

Then $q(x_0) = 0$ and

$$Q_N(x) - q(x) = \frac{1}{n!}\int_{x_0}^{x} (x-t)^n Q_N^{(n+1)}(t)\,dt,$$

Because $q(x)/(x-x_0)$ is a polynomial of degree $\leqslant n-1$ we can write

$$\Delta^2\left(\frac{Q_N(x)}{x-x_0}\right)^{(n-2)}(x,h)=\Delta^2\left(\frac{Q_N(x)-q(x)}{x-x_0}\right)^{(n-2)}(x,h)=$$

$$=\Delta^2\left[\frac{1}{n(n-1)}\sum_{\nu=0}^{n-2}(-1)^{\nu}C_{n-2}^{\nu}\frac{1}{(x-x_0)^{\nu+1}}\int_{x_0}^{x}(x-t)^{\nu+2}Q_N^{(n+1)}(t)\,dt\right](x,h).$$

We consider two cases.

1. $h\geqslant\frac{1}{4}|x-x_0|$. Then for $y=x,\ x+h,\ x+2h$ (3.0.4) yields

$$\frac{1}{|y-x_0|^{\nu+1}}\left|\int_{x_0}^{y}|y-t|^{\nu+2}|Q_N^{(n+1)}(t)|\,dt\right|\underset{n}{\leqslant}|y-x_0|^2\lambda N\underset{n}{\leqslant}\lambda h. \tag{3.0.7}$$

2. $h<\frac{1}{4}|x-x_0|$. Then for $y=x+h,\ x+2h$ we have

$$|y-x_0|\asymp|x-x_0|.$$

For any function ω we obviously have

$$|\Delta^2\omega(x,h)|\leqslant C_0\max_{[x,x+2h]}|\omega''|\cdot h^2. \tag{3.0.8}$$

Hence for the function

$$\omega(y)=\frac{1}{(y-x_0)^{\nu+1}}\int_{x_0}^{y}(y-t)^{\nu+2}Q_N^{(n+1)}(t)\,dt,$$

we have

$$|\omega''(y)|\underset{n}{\leqslant}\lambda N\asymp\lambda/h,\quad y\in[x,x+2h]$$

which together with (3.0.8) implies $|T_2|\underset{n}{\leqslant}\lambda h$. Lemma is proved.●

LEMMA 3.2. Let

$$I\subset\mathbb{R},\ I_0=[-h,h]\subset I,\ f\in Z(I),$$

$$|\Delta^2 f(x,\delta)|\leqslant\lambda\delta,\ x,\ x+h,\ x+2h\in I,\ f(-h)=f(h)=0.$$

Then

$$|f(x)| \lesssim \lambda y \left|\log \frac{y}{h}\right|, \qquad y = \min(|x-h|, |x+h|). \tag{3.0.9}$$

PROOF is a direct consequence of the definition of the class $Z(I)$. ●

LEMMA 3.3. Let

$$I \subset \mathbb{R}, \quad I_0 = [-h, h] \in I, \quad f \in C^{n-1} Z(I),$$

$$|\Delta^2 f^{(n-1)}(x, \delta)| \leq \lambda \delta, \qquad x, x+h, x+2\delta \in I;$$

suppose the points x_ν, $\nu = 0, \ldots, n$ belong to I_0, $x_0 = 0$, $x_n = h$.

Then

a) $$|f(x)| \underset{n}{\lesssim} \lambda(|x|+h)^n \log \frac{|x|+h}{h}, \quad x \in I. \tag{3.0.10}$$

b) if $x_\nu = \frac{\nu h}{n}$, $0 \leq \nu \leq n$ then

$$|f^{(\nu)}(x)| \underset{n}{\lesssim} \lambda(|x|+h)^{n-\nu} \log \frac{|x|+h}{h}, \qquad x \in I, \quad \nu = 0, \ldots, n-1. \tag{3.0.11}$$

PROOF. Let $x_0 \leq x_1 \leq \ldots \leq x_n$

$$F(x) = \frac{f(x)}{(x-x_1)\ldots(x-x_{n-1})}.$$

Lemma 3.1 yields

$$F \in Z(I), \quad |\Delta^2 F(x, \delta)| \underset{n}{\lesssim} \lambda \delta, \qquad x, x+\delta, x+2\delta \in I.$$

Moreover $F(x_0) = F(x_n) = 0$. We apply Lemma 3.2. to the function F and obtain (3.0.10). We prove (3.0.11) by induction in n. For $n = 1$ the result follows from Lemma 3.1. Suppose that the statement of Lemma holds for $C^{n-1} Z$. Let

$$f \in C^n Z(I), \quad |\Delta^2 f^{(n)}(x,\delta)| \leqslant \lambda\delta, \quad f(x_j) = 0, \quad j = 0,1,\ldots,n+1.$$

We put $f_j(x) = f(x)/(x - x_j)$. Lemma 3.1. implies

$$f_j \in C^{n-1} Z(I), \quad |\Delta^2 f_j^{(n-1)}(x,\delta)| \underset{n}{\leqslant} \lambda\delta,$$

$$j = 0,1,\ldots,n+1, \quad x, x+\delta, x+2\delta \in I$$

Hence by part a) we have

$$|f_j(x_j)| \underset{n}{\leqslant} \lambda h^n. \tag{3.0.12}$$

Further

$$f'(x_j) = [(x - x_j) f_j(x)]'\big|_{x_j} = f_j(x_j).$$

Let P be an interpolating polynomial for f' in the points x_j, $j=0,\ldots,n$ then

$$P(x) = \sum_{j=0}^{n} f_j(x_j) \frac{(x - x_0)\ldots(\widehat{x - x_j})\ldots(x - x_n)}{(x_j - x_0)\ldots(x_j - x_n)}.$$

Let $\Phi = f' - P$. Then

$$\Phi \in C^{n-1} Z(I), \ |\Delta^2 \Phi^{(n-1)}(x,\delta)| \leqslant \lambda\delta, \quad [x, x+2\delta] \subset I$$

and $\Phi(x_j) = 0$, $j = 0,\ldots,n$. By induction we write

$$|\Phi^{(\nu)}(x)| \underset{n}{\leqslant} \lambda(|x| + h)^{n-\nu} \log \frac{|x|+h}{h}, \quad 0 \leqslant \nu \leqslant n-1, \tag{3.0.13}$$

and (3.0.12) implies

$$|P^{(\nu)}(x)| \underset{n}{\leqslant} \lambda(|x|+h)^{n-\nu}, \quad 0 \leqslant \nu \leqslant n-1. \tag{3.0.14}$$

Put then in view of $f^{(\nu+1)} = P^{(\nu)} + \Phi^{(\nu)}$ we accomplish the induction step by means of (3.0.13) and (3.0.14). ●

LEMMA 3.4. Let

$$I \subset I \subset \mathbb{R}, \quad |I| \leqslant A|I_0|, \quad f \in C^{n-1} Z(I), \quad M_f(I_0) \geqslant \alpha |I_0|^n.$$

Then

$$M_f(I) \underset{n,A,\alpha,\lambda}{\preccurlyeq} M_f(I_0)$$

where

$$\lambda = \sup_{[x,x+2\delta]\subset I} |\Delta^2 f^{(n-1)}(x,\delta)|\cdot \delta^{-1}.$$

PROOF. Let P, $\deg P \leqslant n$ be the interpolating polynomial for equidistributed points on I_0. Then the assumption and Lemma 3.3 imply

$$M_P(I_0) \leqslant M_f(I_0) + M_{f-P}(I_0) \underset{n}{\preccurlyeq} M_f(I_0) + \lambda|I_0|^n \underset{n,\alpha,\lambda}{\preccurlyeq} M_f(I_0),$$

Hence due to Lemma 3.3 and

$$M_P(I) \underset{A,n}{\preccurlyeq} M_P(I_0)$$

we have

$$M_f(I) \leqslant M_P(I) + M_{f-P}(I) \underset{A,n}{\preccurlyeq} M_P(I_0) + \lambda|I|^n \underset{n,A,\alpha,\lambda}{\preccurlyeq} M_f(I_0). \qquad \bullet$$

LEMMA 3.5. Let

$$f \in C^{n-1}Z(I),\ |\Delta^2 f^{(n-1)}(x,\delta)| \leqslant \delta,\ [x,x+2\delta]\subset I,\ M_f(I) \geqslant \frac{1}{\sigma}|I|^n,\ I_0 \subset I,\ |I_0| = \sigma I.$$

A be a number. Then for $\sigma \leqslant \sigma_n(A)$ we have

$$M_f(I_0) \geqslant A|I_0|^n.$$

PROOF. Let P be the interpolating polynomial of f, $\deg P \leqslant n$, at equidistributed points x_ν, $x_\nu \in I_0$, $x_n - x_0 = |I_0|$. For the function $\varphi = f - P$ we have

$$|\Delta^2 \varphi^{(n-1)}(x,\delta)| \leqslant \delta, \quad [x,x+2\delta]\subset I, \quad \varphi(x_\nu) = 0, \quad 0 \leqslant \nu \leqslant n.$$

Hence Lemma 3.3 yields

$$|\varphi(x)| \leqslant C_{n1}|I|^n \log\frac{|I|}{|I_0|} = C_{n1}|I|^n \log\frac{1}{\sigma}, \quad x \in I. \tag{3.0.15}$$

Now if we would have $M_f(I_0) < A|I_0|^n$ then we could deduce

$$|P(x)| \leqslant \sum_{\nu=0}^{n} |f(x_\nu)| \left| \frac{(x-x_0)\dots\widehat{(x-x_\nu)}\dots(x-x_n)}{(x_\nu - x_0)\dots(x_\nu - x_n)} \right| \leqslant C_{n2} A |I_0|^n \cdot \left(\frac{I}{|I_0|}\right)^n =$$

$$= C_{n2} A |I|^n, \quad x \in I. \qquad (3.0.16)$$

Then (3.0.15) and (3.0.16) would imply

$$\frac{1}{\sigma}|I|^n \leqslant M_f(I) \leqslant M_{f-p}(I) + M_p(I) \leqslant C_{n1}|I|^n \log\frac{1}{\sigma} + AC_{n2}|I|^n,$$

but that is impossible for $\sigma < \sigma_n(A)$, where $\sigma_n(A)$ is the only root of the equation

$$\frac{1}{\sigma_n(A)} = 2(C_{n1} + AC_{n2})(1 + \log\frac{1}{\sigma_n(A)}).$$

Lemma is proved. ●

3.1. Part a) of Theorem 7.

Let u be a conformal mapping which maps $\mathbb{D}$ on Π such that be $\operatorname{Re} u(0) \geqslant c_0 > 0$, $|u(0)| \leqslant c_1$. It is well-known that such a mapping transforms a function $f \in \Lambda^{n-1} Z(\mathbb{D})$ into a function $F = f \circ u^{-1} \in \Lambda^{n-1} Z(\Pi)$. Therefore it is sufficient to prove the natural analogue of (1) for the half-plain and then to return to $\mathbb{D}$ with the help of u^{-1} keeping in mind the fact that the analogue of (1) in the half-plain turns, by the change of variables $\zeta = u^{-1}(z)$ into (1) for $\mathbb{D}$. So we work in the half-plain Π . Since the proof of the part a) of Theorem 7 is similar to that of Theorem 5 our arguments will be brief. Let

$$F \in \Lambda^{n-1} Z(\Pi), \qquad |F(\zeta)| \leqslant 1, \qquad \zeta \in \bar{\Pi},$$

$$|\Delta^2 F^{(n-2)}(x,\delta)| \leqslant \delta, \quad x \in \mathbb{R}, \quad \zeta = x^\circ + iy \in \Pi, \quad |\zeta| \leqslant 10.$$

Suppose first that

$$M_F(\zeta) \overset{def}{=\!=} \max_{[x^\circ - y,\, x^\circ + y]} |F| \geqslant A_n y^n,$$

where the number $A_n > 1$ will be define below. Let $I = [x^0 - y, x^0 + y]$ the points x_ν, $0 \le \nu \le n$, be equidistributed on I, $x_0 = x^0 - y$, $x_n = x_0 + y$, P is the interpolating polynomial such that $P(x_\nu) = F(x_\nu)$, $0 \le \nu \le n$. Then Lemma 3.3 says that

$$|F(x) - P(x)| \le C_n y^n, \quad x \in I.$$

Hence for a sufficiently big A_n we have $M_P(\zeta) \asymp M_F(\zeta)$. Let

$$\zeta_0^* \in I, \quad |F(\zeta_0^*)| = M_F(\zeta), \quad \zeta_\nu^* = \zeta_0^* + i\nu y, \quad 1 \le \nu \le n+1.$$

Then $F \in \Lambda^{n-1} Z(\Pi)$ implies

$$\left|\sum_{\nu=0}^{n+1} (-1)^\nu C_{n+1}^\nu F(\zeta_\nu^*)\right| \le C_{n3}\, y^n,$$

and similarly to (1.1.2) - (1.1.3) we find

$$|F(\zeta_\nu^*)| \underset{n}{\lesssim} M_F(\zeta) e^{-C_{n4} B}, \quad 1 \le \nu \le n+1,$$

where

$$B = \int_{\mathbb{R}} \left|\log\left|\frac{F(x)}{P(x)}\right|\right| \frac{y}{|x-\zeta|^2}\, dx,$$

and then, as in (1.1.4) - (1.1.5) we obtain the inequality $B \le C_{n5}$ for sufficiently big A_n . The case $M_f(\xi) \ge y^n$ follows from the previous case with the help of Lemma 3.4 in a manner like (1.1.6) - (1.1.7).

3.2. The first crucial estimate.

We recall that in the remainder part of the proof we deal with the half-plain Π instead of $\mathbb{D}$.

LEMMA 3.6. Let

$$F \in C^{n-1} Z(\mathbb{R}), \qquad |\Delta^2 F^{(n-1)}(x,\delta)| \le \delta.$$

Suppose that F satisfies the assumption of the part b) of Theorem 7 on the line $\mathbb{R}$ and let $\zeta = x_0 + i y_0 \in \Pi$, $|\zeta| \le 10$, the points x_ν^*,

$0 \leq \nu \leq n$, being equidistributed on $[x_0 - y_0, x_0 + y_0]$, $x_0^* = x_0 - y_0$, $x_n^* = x_0 + y_0$, P be an interpolating polynomial, $P(x_\nu^*) = F(x_\nu^*)$, and let

$$L = D(x_0, v, V), \quad v \geq y, \quad \frac{4}{3} \leq \frac{V}{v} \leq \frac{5}{2},$$
$$I = \mathbb{R} \cap L, \quad L^0 = D(x_0, q_n'' v, q_n' V)$$

$\lambda_1, \dots, \lambda_m$, $1 \leq m \leq n$ being the zeros of P in the annulus L ; suppose next that $\zeta_\nu \in L_0$, $1 \leq \nu \leq m$. We denote

$$r(\lambda) = \prod_{\nu=1}^{m} (\lambda - \lambda_\nu)$$

$$Q(\lambda) \stackrel{def}{=} {}_E r(\lambda) = \exp \frac{1}{\pi i} \int_{\mathbb{R}} \log |r(x)| \frac{dx}{x - \lambda}, \ \lambda \in \Pi,$$

$$\sigma = \frac{V^n \log \frac{V}{y_0}}{\max\limits_{D(x_0, V)} |P|},$$

$$\varphi(\lambda) \stackrel{def}{=} Q(\lambda) \exp \frac{1}{\pi i} \int_I \log \left| \frac{F(x)}{P(x)} \right| \frac{dx}{x - \lambda}. \tag{3.2.0}$$

Then there exists $\sigma_n > 0$ such that if $\sigma \leq \sigma_n$ then

$$|\varphi^{(\nu)}(\zeta)| \underset{\nu, n, F}{\leq} V^{m-\nu}, \ \nu \geq 0, \tag{3.2.1}$$

$$|\varphi^{(\nu)}(\zeta)| \underset{\nu, n, F}{\leq} \sigma V^{m-\nu}, \quad \nu \geq m+1; \tag{3.2.2}$$

b) The estimates (3.2.1) and (3.2.2) hold in the case $L = \mathcal{D}(x_0, V)$, $2y_0 \leq V \leq 3y_0$ and the polynomial P has no zeros in the annulus $D(x_0, q_n' V, V)$.

PROOF. We first prove the estimate (3.2.1). If $\sigma \leq \sigma_{n1}$ then

$$M_P([x_0 - V, x_0 + V]) \asymp M_F([x_0 - V, x_0 + V])$$

and hence for $\tilde{\zeta} = x_0 + iV$ and for a suitable choice of σ_{n1} we have $M_F(\tilde{\zeta}) \geq V^n$ which together with the assumption b) of Theorem

7 yields

$$\int_I |\log|\frac{F(t)}{P(t)}||\,dt \underset{F,n}{\preccurlyeq} V,$$

and then

$$\int_I |\log|\frac{F(t)}{P(t)}||\,\frac{dt}{|t-\zeta|^{\nu}} \underset{\nu,F,n}{\preccurlyeq} V^{1-\nu}, \quad \nu\geqslant 0. \tag{3.2.3}$$

Differentiating φ and bearing in mind (3.2.3) we obtain (3.2.1). Let us proceed now to (3.2.2). There are two different cases.

1. $m<n$. Let $R=P/r$, $R_0=|R(\lambda)|$, $H=F-P$.

Lemma 3.3 yields

$$|H^{(\nu)}(x)| \underset{n}{\preccurlyeq} V^{n-\nu}\log\frac{V}{y_0}, \quad x\in[x_0-V,\, x_0+V],\ 0\leqslant\nu\leqslant n-1 \tag{3.2.4}$$

and in view of $\max\limits_{D(x_0,V)}|P| \underset{n}{\asymp} R_0V^m$ we can rewrite (3.2.4) in the following form:

$$|H^{(\nu)}(x)| \underset{n}{\preccurlyeq} \frac{V^n\log\frac{V}{y_0}}{\max\limits_{D(x_0,V)}|P|}R_0V^{m-\nu} = \sigma R_0V^{m-\nu}.$$

Then since $R\neq 0$ in L we find

$$\left|\left(\frac{H}{R}\right)^{(\nu)}(x)\right| \underset{n}{\preccurlyeq} \sigma V^{m-n}, \quad 0\leqslant\nu\leqslant n-1, \quad x\in[x_0-V,\, x_0+V]. \tag{3.2.5}$$

Let $h=H/R$. The estimated (3.2.5) allow us to apply to the function

$$\varphi(\lambda) = {}_E r(\lambda)\cdot\exp\frac{1}{\pi i}\int_I \log\left|1-\frac{h(t)}{r(t)}\right|\frac{dt}{t-\lambda}$$

an analogue of Lemma 1.9 for the half-plain.

2. $m=n$. It is the most difficult case. We notice first the following circumstance. Let $x_0^*=x_0-V$, $x_\nu^*\in[x_0-q'_{n+1}V,\, x_0+q'_{n+1}V]$, $1\leqslant\nu\leqslant n$. Then $x_\nu^*-x_0^*\geqslant q'_{n+1}V$ and if P_{X^*} is the interpolating polynomial of F for the points $x_0^*, x_1^*, \dots, x_n^*$ then by Lemma 3.3 we have

$$|F(x)-P_{X^*}(x)| \leqslant C_{n1} V^n \log \frac{V}{y_0}, \quad x \in [x_0-V, x_0+V].$$

Hence for any such polynomial P we obtain the inequalities

$$|P(x)-P_{X^*}(x)| \leqslant C_{n2} V^n \log \frac{V}{y_0}, \quad x \in [x_0-V, x_0+V]. \tag{3.2.6}$$

The location of zeros of P together with (3.2.6) yields

$$\left| \frac{P(\lambda)-P_{X^*}(\lambda)}{P(\lambda)} \right| \leqslant C_{n3}\, \sigma, \quad \lambda \in \partial \mathbb{D}(x_0, q''_{n+1} v, q'_{n+1} V). \tag{3.2.7}$$

Hence if $\sigma \leqslant \sigma_n$ where

$$C_{n3}\, \sigma_n \leqslant \frac{1}{2} \tag{3.2.8}$$

the polynomial P_X has in the annulus $D(x_0, q''_{n+1} v, q'_n V)$ n zeros like the polynomial P . We now change the polynomial P in (3.2.0) to the polynomial P_{X^*} where $X^* = (x_1^*, \dots, x_n^*)$ is suitable chosen.

Let $\zeta_1(X^*), \dots, \zeta_n(X^*)$ be all the zeros of P_{X^*} numbered in such a way that $\xi_n(X^*) \overset{def}{=} \operatorname{Re} \zeta_\nu(X^*) \leqslant \xi_{\nu+1}(X)$, $1 \leqslant \nu \leqslant n-1$. Then the vector $\xi(X^*) \overset{def}{=} (\xi_1(X^*), \dots \xi_n(X^*))$ depends continuously on $X^* = (x_1^*, \dots, x_n^*)$. Due to the facts that $x_\nu^* \in [x_0 - q'_{n+1} V, x_0 + q'_{n+1} V]$, $1 \leqslant \nu \leqslant n$ and that all of the zeros of P_{X^*} lie in $D(x_0, q''_{n+1} v, q'_{n+1} V)$, i.e. $\xi_\nu(X^*) \in [x_0 - q'_{n+1} V, x_0 + q'_{n+1} V]$, $1 \leqslant \nu \leqslant n$ the Brauer theorem on fixed point yields that the mapping $X^* \longrightarrow \xi(X^*)$ has a fix point, i.e. there exists a point $x_1^+, \dots, x_n^+ \in [x_0 - q'_{n+1} V, x_0 + q'_{n+1} V]$ for which $\xi_\nu(X^+) = x_\nu^+$, $1 \leqslant \nu \leqslant n$. Let P_+ be the polynomial corresponding to the vector $(x_1^+, \dots, x_n^+)$, a and a_+ being the senior coefficient of P and P_+ , $r_+ = P_+/a_+$, $Q_+ = {}_E r_+$. Then

$$\varphi(\lambda) =$$

$$= \left[Q_+(\lambda)\cdot \exp\frac{1}{\pi i}\int\limits_I \log\left|\frac{F(t)}{P_+(t)}\right|\frac{dt}{t-\lambda}\right]\left[\frac{Q(\lambda)}{Q_+(\lambda)}\cdot \exp\frac{1}{\pi i}\int\limits_I \log\left|\frac{P_+(t)}{P(t)}\right|\frac{dt}{t-\lambda}\right] \stackrel{def}{=\!=}$$

$$\stackrel{def}{=\!=} \varphi_+(\lambda)\, q(\lambda). \tag{3.2.9}$$

We conclude from (3.2.7) that

$$\left|\frac{a_+}{a} - 1\right| \underset{n}{\lessdot} \sigma, \tag{3.2.9'}$$

which together with (3.2.7) allows us to apply Lemma 1.8 to q in the case $\sigma \leqslant \sigma_n$ and to obtain

$$|q(\zeta)| \lessdot 1, \tag{3.2.10}$$

$$|q^{(\nu)}(\zeta)| \underset{\nu,n}{\lessdot} \sigma V^{-\nu}, \quad \nu \geqslant 1. \tag{3.2.11}$$

We are going to prove only the estimates (3.2.2) for φ_+. First we rewrite the expression for σ. We have

$$\max_{D(x_0,V)} |P| \underset{n}{\asymp} |a|\, V^n,$$

$$\sigma = \frac{V^n \log\frac{V}{y_0}}{\max\limits_{D(x_0,V)}|P|} \asymp \frac{\log\frac{V}{y_0}}{|a|}$$

and since $|a| \asymp |a_+|$ by (3.2.9') we obtain

$$\sigma \underset{n}{\asymp} \frac{\log\frac{V}{y_0}}{|a_+|} \geqslant \frac{1}{|a_+|}. \tag{3.2.12}$$

Let

$$h_+ = F - P_+,$$

$$g(x) = \frac{h_+(x)}{(x-x_1^+)\dots(x-x_{n-1}^+)}, \quad n \geq 2,$$

$$g(x) = h_+(x), \quad n = 1.$$

Lemma 3.2 yields $g \in Z([x_0-V, x_0+V]), |\Delta^2 g(x,\delta)| \underset{n}{\leq} \delta, [x, x+2\delta] \subset [x_0-V, x_0+V]$ but $g(x_n^+) = 0$. Hence

$$|g(x)| \underset{n}{\leq} |x - x_n^+| \log \frac{4V}{|x - x_n^+|}, \quad x \in [x_0 - V, x_0 + V]. \tag{3.2.13}$$

But

$$\frac{F(x)}{P_+(x)} = 1 + \frac{1}{a_+} \cdot \frac{(x-x_1^+)\dots(x-x_{n-1}^+)}{(x-\zeta_1^+)\dots(x-\zeta_{n-1}^+)} \cdot \frac{g(x)}{x - \zeta_n^+},$$

where ζ_n^+, $1 \leq \nu \leq n$ are zeros of P_+. Since $\operatorname{Re} \zeta_\nu^+ = x_\nu^+$, $1 \leq \nu \leq n$, we obtain the following simple but decisive property: for $x \in R$

$$\left| \frac{x - x_\nu^+}{x - \zeta_\nu^+} \right| \leq 1, \qquad 1 \leq \nu \leq n$$

and then (3.2.12) and (3.2.13) imply

$$\left| \frac{1}{a_+} \cdot \frac{(x-x_1^+)\dots(x-x_{n-1}^+)}{(x-\zeta_1^+)\dots(x-\zeta_{n-1}^+)} \cdot \frac{g(x)}{x-\zeta_n^+} \right| \leq \frac{1}{|a_+|} \left| \frac{x - x_n^+}{x - \zeta_n^+} \right| \log \frac{4V}{|x - x_n^+|} \lessapprox$$

$$\underset{n}{\lessapprox} \sigma \log \frac{4V}{|x - x_n^+|}, \quad x \in [x_0 - V, x_0 + V]. \tag{3.2.14}$$

If $\sigma \leq \sigma_{n3}$ then Lemma 3.3 and the inequality

$$\max_{[x_0 - V, x_0 + V]} |P| \geq C_{n3} \frac{1}{\sigma} V^n \log \frac{V}{y_0}$$

yields

$$M_F([(x_0 - V, x_0 + V)]) \geq C_{n4} \frac{1}{\sigma} V^n \log \frac{V}{y_0} \geq C_{n4} \frac{1}{\sigma} V^n. \tag{3.2.15}$$

We choose $\sigma \leq \sigma_{n4}$ where $C_{n4}/\sigma_{n4} \geq \sigma_n^{-1}(1)$ (Lemma 4.5). Then for $\sigma \leq \sigma_{n5}$ making use of (3.2.15) for $|x-x_n^+| \geq \frac{\sigma}{4} V$ we obtain the estimate

$$(3.2.14) \leq C_{n5}\, \sigma \log \frac{16}{C_{n4}\sigma} \leq \frac{1}{2} \tag{3.2.16}$$

i.e. for $\sigma \leq \sigma_{n5}$ and $|x-x_n^+| \geq \frac{\sigma}{4} V$ we can write

$$\left|\log\left|\frac{F(x)}{P_+(x)}\right|\right| \underset{n}{\leq} \sigma \log \frac{4V}{|x-x_n^+|}. \tag{3.2.17}$$

Further in view of the inequality $C_{n4}/\sigma_{n4} \geq \sigma_n^{-1}$ (1) (Lemma 3.5) Lemma 3.5 implies

$$M_F([x_n^+ - \sigma V, x_n^+ + \sigma V]) \geq (\sigma V)^n$$

and then the assumption b) of Theorem 7 which figures in the assumption of the lemma yields

$$\int\limits_{[x_n^+-\sigma V,\, x_n^++\sigma V]} \log\left|\frac{F(x)}{M_F(\tau)}\right| dx \underset{n}{\leq} \sigma V, \tag{3.2.18}$$

where $\tau = [x_n^+ - \sigma V, x_n^+ + \sigma V]$.

For $\sigma \leq \sigma_n$, where σ_n is chosen suitable, Lemma 3.5 with $A=1$ can be applied to the segment $\tau_0 = [x_n^+ + \frac{\sigma}{2} V, x_n^+ + \sigma V]$. Then for $x \in \tau_0$, we have

$$|F(x)| \asymp |P_+(x)|, \tag{3.2.19}$$

and in particular

$$M_F(\tau_0) \asymp M_{P_+}(\tau_0). \tag{3.2.20}$$

The variant of Lemma 3.5 in the case of the segment τ_0 with $A=1$ implies $M_F(\tau_0) \geq |\tau_0|^n$ and then Lemma 3.4 yields

$M_F(\tau) \underset{n}{\asymp} M_F(\tau_0)$ hence noticing (3.2.20) we have

$$M_{P_+}(\tau) \underset{n}{\asymp} M_{P_+}(\tau_0) \asymp M_F(\tau_0) \asymp M_F(\tau). \tag{3.2.21}$$

As a consequence of (3.2.21), (3.2.18) and Lemma 1.3 we get

$$\int_{\tau} \|\log\cdot\| dx \leq \int_{\tau} \left|\log\left|\frac{F(x)}{M_F(\tau)}\right|\right| dx + \int_{\tau} \left|\log\left|\frac{P_+(x)}{M_F(\tau)}\right|\right| dx \underset{n}{\leqslant}$$

$$\underset{n}{\leqslant} \sigma V + \int_{\tau} \left|\log\left|\frac{P_+(x)}{M_{P_+}(\tau)}\right|\right| dx + 2\sigma V \left|\log\frac{M_{P_+}(\tau)}{M_F(\tau)}\right| \underset{n}{\leqslant} \sigma V. \tag{3.2.22}$$

But (3.2.22) and (3.2.14) imply

$$\int_I \left|\log\left|\frac{F(x)}{P_+(x)}\right|\right| dx = \int_{\tau} + \int_{I\setminus\tau} \underset{n}{\leqslant} \sigma V + \sigma \int_I \log\frac{4V}{|x-x_n^+|} dx \underset{n}{\leqslant}$$

$$\underset{n}{\leqslant} \sigma V + 2\sigma \int_0^V \log\frac{4V}{t} dt \underset{n}{\leqslant} \sigma V, \tag{3.2.23}$$

and then due to (3.2.23)

$$\int_I \left|\log\left|\frac{F(x)}{P_+(x)}\right|\right| \frac{d(x)}{|x-\zeta|^{\nu}} \underset{n,\nu}{\leqslant} \sigma V^{1-\nu}, \quad \nu \geqslant 0. \tag{3.2.24}$$

Finally, if we differentiate the function φ_+, $\nu \geqslant n+1$ times then we differentiate the exponent at least once (the terms in which we differentiate only the polynomial Q_+ vanish identially since $\deg Q_+ = n$). Hence (3.2.24) provides the inequality required. ●

3.3. The second crucial estimate.

LEMMA 3.7. a) Let $\xi \in \Pi$, F and a polynomial P be taken from Lemma 3.2. Suppose that P has no zeros in the annulus $L_0 = D(x_0, q_n' v, q_n'' V)$, where $v \geqslant y_0$, $V/v \geqslant 4/3$ and has $m, m \leqslant n$ zeros in $D(x_0, v)$. Denote

$$\sigma_\ell = \frac{\ell^n \log \frac{2\ell}{y_0}}{\max\limits_{D(x_0,\ell)} |P|}$$

$$I = D(x_0, v, V) \cap \mathbb{R},$$

$$\varphi(\lambda) \overset{\text{def}}{=} \exp \frac{1}{\pi i} \int\limits_I \log \left| \frac{F(x)}{P(x)} \right| \frac{dx}{x - \lambda}.$$

Then there exists $\tilde{\sigma}_n > 0$ such that of $\sigma_V \leqslant \tilde{\sigma}_n$ then

$$|\varphi^{(\nu)}(\zeta)| \underset{\nu,n}{\lesssim} v^{-\nu}, \quad \nu \geqslant 0 \tag{3.3.1}$$

$$|\varphi^{(\nu)}(\zeta)| \underset{\nu,n}{\lesssim} \sigma_v v^{-\nu} + \sigma_V V^{-\nu}, \quad \nu \geqslant 1, \quad \nu \neq n-m, \tag{3.3.2}$$

$$|\varphi^{(n-m)}(\zeta)| \underset{\nu,n}{\lesssim} \sigma_V V^{-(n-m)} \log \frac{V}{v}. \tag{3.3.3}$$

b) The estimates (3.3.1) - (3.3.3) hold if

$$D(x_0, V), \quad I = [x_0 - V, x_0 + V], \quad V \geqslant 2y_0$$

P has no zeros in $D(x_0, q_n'' V)$, where $v = y_0$.

PROOF. We restrict ourselves only with the proof of the more difficult part a). Let $\lambda_1, \dots, \lambda_m$ be all the zeros of P in $\mathcal{D}(x_0, v)$,

$$Q(\lambda) = P(\lambda) / \prod_{\nu=1}^{m} (\lambda - \lambda_\nu), \quad Q = |Q(x_0)|,$$

since by Lemma 3.3 we have

$$|F(x) - P(x)| \underset{n}{\lesssim} |x - x_0|^n \log \frac{2|x - x_0|}{y_0}, \quad x \in I,$$

so similarly to the steps 1.3 and 2.3 for $\sigma_V \leqslant \tilde{\sigma}_n, \psi = \log \varphi$ and $r \geqslant 1$ we have

$$|\varphi^{(r)}(\zeta)| \underset{r,n}{\preccurlyeq} \int_I \frac{|x-x_0|^n \log\frac{2|x-x_0|}{y_0}}{Q\,|x-x_0|^m} \cdot \frac{dx}{|x-\zeta|^{r+1}} \underset{r,n}{\asymp}$$

$$\underset{r,n}{\asymp} \frac{1}{Q}\int_v^V t^{n-m-r-1} \log\frac{2t}{y_0}\, dt \preccurlyeq$$

$$\preccurlyeq \begin{cases} \frac{1}{Q} V^{n-m-r} \log\frac{V}{y_0}, & n-m-r>0, \\ \frac{1}{Q} \log\frac{V}{y_0} \cdot \log\frac{V}{v}, & n-m-r=0, \\ \frac{1}{Q} v^{n-m-r} \log\frac{2v}{y_0}, & n-m-r<0. \end{cases} \tag{3.3.4}$$

The estimates (3.3.1) are analogous to (2.3.3) and we deal only with the proof of (3.3.2) and (3.3.3). Since

$$\varphi^{(\nu)} = \sum_{\substack{r_1+\dots+r_\ell=\nu \\ r_j>0}} \varphi \cdot C_\nu^{r_1,\dots,r_\ell}\, \psi^{(r_1)} \dots \psi^{(r_\ell)},$$

it is sufficient to estimate each term of the form $t_{r_1,\dots,r_\ell} = \psi^{(r_1)}(\zeta)\dots\psi^{(r_\ell)}(\zeta)$.

CASE 1. For all $j=1,\dots,\ell$, $r_j<n-m$ or for all $j=1,\dots,\ell$, $r_j>n-m$. Then implies

$$t_{r_1,\dots,r_\ell} \underset{r,n}{\preccurlyeq} \frac{1}{Q^\ell} w^{(n-m)\ell-\nu} \cdot \left(\log\frac{2w}{y_0}\right)^\ell =$$

$$= \frac{w^n \log\frac{2w}{y_0}}{Q w^m} \cdot w^{-\nu} \left(\frac{w^n \log\frac{2w}{y_0}}{Q w^m}\right)^{\ell-1} \underset{\nu,n}{\preccurlyeq} \sigma_v v^{-\nu} + \sigma_V V^{-\nu}, \tag{3.3.5}$$

where $w=V$ $r_j<n-m$, $w=v$ $r_j>n-m$.

CASE 2. $r_1,\dots,r_K<n-m$, $r_{K+1}=\dots=r_\ell=n-m$, $\nu>n-m$. Let

$$\nu_1 = \sum_1^k r_j, \qquad \nu_0 = (\ell-K)(n-m).$$

We notice that in our situation $n>m$. Then

$$t_{r_1,\dots,r_\ell} \underset{\nu,n}{\preccurlyeq} \frac{1}{Q^\ell} V^{(n-m)k-\nu_1} \left(\log\frac{V}{y_0}\right)^\ell \left(\log\frac{V}{v}\right)^{\ell-K} =$$

$$= \frac{1}{Q} v^{n-m-\nu} \log\frac{2v}{y_0} \cdot \frac{1}{Q^{\ell-1}} \left(\frac{v}{V}\right)^{\nu_1} V^{k(n-m)} \left(\log\frac{V}{y_0}\right)^{\ell} \cdot \left(\log\frac{V}{v}\right)^{\ell-k} \cdot \frac{v^{(n-m)(\ell-k-1)}}{\log\frac{2v}{y_0}} .$$

If $\ell - k = 1$ then $\nu_1 > 0$ hence

$$\frac{1}{Q^{\ell-1}} \left(\frac{v}{V}\right)^{\nu_1} V^{(n-m)(\ell-1)} \left(\log\frac{V}{y_0}\right)^{\ell} \frac{\log\frac{V}{v}}{\log\frac{2v}{y_0}} =$$

$$= \left(\frac{V^{n-m}\log\frac{V}{y_0}}{Q}\right)^{\ell-1} \cdot \left(\frac{v}{V}\right)^{\nu_1} \cdot \log\frac{V}{v} \cdot \frac{\log\frac{V}{y_0}}{\log\frac{2v}{y_0}} \underset{\nu,n}{\ll}$$

$$\underset{\nu,n}{\ll} \left[\left(\frac{2v/y_0}{V/y_0}\right)^{\nu_1/2} \cdot \frac{\log\frac{V}{y_0}}{\log\frac{2v}{y_0}}\right] \cdot \left[\left(\frac{2v/y_0}{V/y_0}\right)^{\nu_1/2} \log\frac{V/y_0}{2v/y_0}\right] \underset{\nu,n}{\ll} 1,$$

and then

$$t_{r_1,\dots,r_\ell} \underset{\nu,n}{\ll} \sigma_v v^{-\sigma}, \tag{3.3.6}$$

But if $\ell - k > 1$ then

$$\frac{1}{Q^{\ell-1}} \left(\frac{v}{V}\right)^{\nu_1} V^{(n-m)k} \left(\log\frac{V}{y_0}\right)^{\ell} \left(\log\frac{V}{v}\right)^{\ell-k} \frac{v^{(n-m)(\ell-k-1)}}{\log\frac{2v}{y_0}} \leqslant$$

$$\leqslant \frac{1}{Q^{\ell-1}} V^{(n-m)k} \left(\log\frac{V}{y_0}\right)^{\ell} \left[\left(\log\frac{V}{v}\right)^{\ell-k} \frac{v^{(n-m)(\ell-k-1)}}{\log\frac{2v}{y_0}}\right] \underset{\nu,n}{\ll}$$

$$\underset{\nu,n}{\ll} \frac{1}{Q^{\ell-1}} V^{k(n-m)} \left(\log\frac{V}{y_0}\right)^{\ell} \frac{V^{(n-m)(\ell-k-1)}}{\log\frac{V}{y_0}} = \left(\frac{V^{n-m}\log\frac{V}{y_0}}{Q}\right)^{\ell-1} \underset{\nu,n}{\ll} 1,$$

and again

$$t_{r_1,\dots,r_\ell} \underset{\nu,n}{\preccurlyeq} \sigma_v v^{-\nu}. \tag{3.3.7}$$

CASE 3. $\ell = 1$, $r = n-m$. It is sufficient to apply (3.3.4).

CASE 4. $0 < k + L < \ell$ and

$$r_1, \dots, r_k < n-m; \quad r_{k+1}, \dots, r_{k+L} = n-m; \quad r_{k+L+1}, \dots, r_\ell > n-m.$$

Then again we have $n > m$ because all $r_j \geq 1$.

Let

$$\nu_1 = \sum_1^k r_j, \quad \nu_0 = (n-m)L, \quad \nu_2 = \sum_{k+L+1}^{\ell} r_j.$$

Then $\nu_2 > 0$, $\nu_1 + \nu_0 + \nu_2 = \nu$. The inequalities (3.3.4) yield

$$t_{r_1,\dots,r_\ell} \preccurlyeq \frac{1}{Q^\ell} V^{(n-m)k-\nu_1} \cdot \left(\log \frac{V}{y_0}\right)^k \cdot \left(\log \frac{V}{y_0}\right)^L \cdot \left(\log \frac{V}{v}\right)^L \times$$

$$\times v^{(n-m)(\ell-k-L)-\nu_2} \cdot \left(\log \frac{2v}{y_0}\right)^{\ell-k-L} =$$

$$= \left(\frac{1}{Q^\ell} v^{n-m-\nu} \log \frac{2v}{y_0}\right) \cdot \left[V^{(n-m)k-\nu_1} \cdot \left(\log \frac{V}{y_0}\right)^{k+1} \cdot \left(v^{n-m} \log \frac{2v}{y_0}\right)^{\ell-k-L-1} \cdot v^{\nu_1+\nu_0} \left(\log \frac{V}{v}\right)^L \right] \underset{\nu,n}{\preccurlyeq}$$

$$\underset{\nu,n}{\preccurlyeq} \left(\frac{1}{Q} v^{n-m-\nu} \log \frac{2v}{y_0}\right) \cdot \frac{1}{Q^{\ell-1}} \left[V^{(n-m)k-\nu} \cdot \left(\log \frac{V}{y_0}\right)^{k+L} \left(V^{n-m} \log \frac{V}{y_0}\right)^{\ell-k-L-1} \cdot V^{\nu_1+\nu_0} \right] \underset{\nu,n}{\preccurlyeq}$$

$$\underset{\nu,n}{\preccurlyeq} \sigma_v v^{-\nu} \left(\frac{V^{n-m} \log \frac{V}{y_0}}{Q} \right)^{\ell-1} \underset{\nu,n}{\preccurlyeq} \sigma_v v^{-\nu}. \tag{3.3.8}$$

The estimates (3.3.5) - (3.3.8) finish the proof of the Lemma. ●

3.4. Proof of Theorem 7 - part b).

Let $F \in C^{n-1} Z(\partial \mathbb{D})$, F satisfy the assumption b) of Theorem 7 in the disc $\mathbb{D}$, $f = {}_e F$. We are to prove only the estimate

$$|f^{(n+1)}(z)| \underset{f}{\preccurlyeq} (1-|z|)^{-1}, \quad z \in \mathbb{D},$$

because it is well-known that it yields $f \in \Lambda^{n-1} Z(\mathbb{D})$. Suppose that the function u maps the upper half-plain Π onto $\mathbb{D}$ conformally, $G = F \circ u$, $g = f \circ u$. Then $G \in C^{n-1} Z(\mathbb{R})$, $g =_E G$, and the function G satisfies the assumption b) of Theorem 7 in Π ; moreover, the estimate

$$|f^{(n+1)}(z)| \underset{f}{\lesssim} (1-|z|)^{-1}$$

is equivalent to the estimates

$$|g^{(n+1)}(\zeta)| \underset{g}{\lesssim} y^{-1}, \quad \zeta = x+iy \in \Pi, \quad |\zeta| \leqslant 10$$

obtained for ten "equidistributed" mappings . Bearing it in mind, we intend to prove the inequality

$$|f^{(n+1)}(\zeta)| \underset{f}{\lesssim} \frac{1}{y}, \quad \zeta = x+iy \in \Pi, \quad |\zeta| \leqslant 10, \tag{3.4.1}$$

assuming that $f =_E F$, $F \in C^{n-1} Z(\mathbb{R})$, F satisfies condition b) of Theorem 7 in Π , $|\Delta^2 F^{(n-1)}(x,\delta)| \leqslant \delta$. We consider two possibilities

1. $M_F(\zeta) \leqslant A y^n$, $A = A_n(F)$ is a number, which we define below in the case 2.

Let P be the interpolating polynomial of F for the points $x - \frac{\nu}{n} y$, $\nu = 0, 1, \ldots, n$, $\zeta = x+iy$. Then Lemma 3.3 yields

$$|F(t) - P(t)| \underset{n}{\lesssim} y^n, \quad t \in [x-y, x+y],$$

$$|F(t) - P(t)| \underset{n}{\lesssim} |t-x|^n \log\left(\frac{2|t-x|}{y}\right), \quad t \in [x-y, x+y]. \tag{3.4.2}$$

Hence $M_P(\zeta) \underset{n,A}{\lesssim} y^n$. Now the Cauchy formula gives

$$|f^{(n+1)}(\zeta)| = \left| \frac{(n+1)!}{2\pi i} \int\limits_{|\lambda-\zeta| = \frac{y}{2}} \frac{f(\lambda)\, d\lambda}{(\lambda-\zeta)^{n+2}} \right| \underset{n}{\lesssim} y^{n-1} \max_{\partial D(\zeta, \frac{y}{2})} |f(\lambda)|,$$

and then using Lemma 1.2 and (3.4.2) we obtain

$$|f(\lambda)| \underset{n}{\lesssim} \exp\frac{1}{\pi}\int_{\mathbb{R}} \log[|P(\tau)| + |\tau-x|^n \log\left(\frac{2|\tau-x|+y}{y}\right)] \frac{\eta\, d\tau}{|\tau-\lambda|^2} \underset{n}{\lesssim}$$

$$\underset{n}{\lesssim} \exp\frac{1}{\pi}\int_{\mathbb{R}} \log[M_P(\zeta)\left(\frac{|\tau-x|+y}{y}\right)^n + |\tau-x|^n \log\frac{2(\tau-x)+y}{y}] \frac{\eta\, d\tau}{|\tau-\lambda|^2} \underset{n,A}{\lesssim}$$

$$\underset{n,A}{\lesssim} \exp\frac{1}{\pi}\int_{\mathbb{R}} \log[y^n\cdot\left(\frac{|\tau-x|+y}{y}\right)^n + |\tau-x|^n\cdot \frac{|\tau-x|+y}{y}] \frac{\eta\, d\tau}{|\tau-\lambda|^2} \underset{n,A}{\lesssim}$$

$$\underset{n,A}{\lesssim} y^n, \qquad \lambda = \xi + i\eta \in \partial D(\zeta, \tfrac{y}{2})$$

and then

$$|f^{(n+1)}(\zeta)| \underset{n,A}{\lesssim} \frac{1}{y}. \tag{3.4.3}$$

2. $M_F(\zeta) \geqslant Ay^n$. We now calculate the value of $A = A_n(F)$. Let P be the interpolating polynomial F for the points $x + \frac{\nu}{n}y$, $\nu = 0,1,\ldots,n$. Lemma 3.3 yields

$$|P(t) - F(t)| \leqslant C_{n1} y^n, \quad t \in [x-y, x+y]. \tag{3.4.4}$$

Hence for $A \geqslant 2C_{n1}$ we have $M_P(\zeta) \geqslant \frac{1}{2}Ay^n$. Let $A = 2\cdot 100^{n+1}\tilde{A}C_{n2}$ where

$$\tilde{A} = \max\left(C_{n1}, \frac{1}{\sigma_n(L.3.6)}, \frac{1}{\tilde{\sigma}_n(L.3.7)} A_{(3.4.10)}(F,n)\right) \tag{3.4.5}$$

and C_{n2} be the constant in the inequality

$$\max_{[x-6y,\,x+6y]} |P| \geqslant C_{n2}\cdot 4^n \max_{[x-y,\,x+y]} P, \quad \deg P \leqslant n. \tag{3.4.6}$$

Let

$$s \overset{\text{def}}{=} \max\{\delta:\ \text{for any } \tau,\ 0<\tau\leqslant\delta,\ \max_{D(x,\tau)}|P|\geqslant \tilde{A}(\tau^n+y^n)\log\frac{\tau+6y}{y}\}.$$

In view of our choice of $\tilde{A}$, the assumption b) of Theorem 7 and (3.4.6), we find that $s\geqslant 6y$. Let Ξ be the zero-set of P. For Ξ, for the point x and for the value $\rho=y$ we proceed as in $(1.6.8_1)$ - $(1.6.8_4)$. We choose as in § 1 the number s_0, $\frac{s}{3}\leqslant s_0\leqslant s$ and obtain the set of annuli $\mathcal{L}_0=\{L_j\}_{j=0}^{N+1}$, $N\leqslant 2n+1$, where

$$[\ L_0=D(x,V_0),\quad L_j=D(x,V_{j-1},V_j),\quad 1\leqslant j\leqslant N$$

satisfying $(1.6.8_1)$- $(1.6.8_4)$ for $\rho=y$. It is possible since $s_0\geqslant 2y$. The choice of the numbers s and $\tilde{A}$ permits us to apply Lemma 3.6 to every annulus L_j, $0\leqslant j\leqslant N$ which intersects with Ξ. We apply Lemma 3.7 to the remaining annuli. For the polynomial P, the set of annuli $\{L_j\}_{j=0}^{N+1}$ and for the outer function $f={}_EF$ the factorization of the type (1.6.8) is valid:

$$f(\lambda)=\prod_{j=0}^{N+1}\varphi_j(\lambda), \tag{3.4.7}$$

where the function φ_j corresponds to the annulus L_j. Let m_j, $m_j\geqslant 0$ be the number of zeros of P in the annulus L_j, $0\leqslant j\leqslant N$. We let $m_{N+1}=0$. We recall the estimates for $\varphi_j^{(\nu)}(\xi)$ obtained earlier in 3.3.2 and 3.3. Let

$$V_{-1}\overset{\text{def}}{=}y,\qquad l_j=\sum_{K=0}^{j}m_K,\quad 0\leqslant j\leqslant N,$$

$$\sigma_j=\left(V_j^n\log\frac{4V_j}{y}\right)\Big/\max_{D(x,V_j)}|P|,\quad j=-1,0,1,\dots,N.$$

We have

$$\left.\begin{aligned}&|\varphi_j^{(\nu)}(\zeta)| \underset{F,\nu}{\lessapprox} V_{j-1}^{-\nu},\ \nu \geqslant 0,\\&|\varphi_j^{(\nu)}(\zeta)| \underset{F,\nu}{\lessapprox} \sigma_{j-1} V_{j-1}^{-\nu} + \sigma_j V_j^{-\nu},\ \nu \geqslant 1,\ \nu \neq n-\ell_j\\&|\varphi_j^{(\ell_j)}(\zeta)| \underset{F,n}{\lessapprox} \sigma_j V_j^{-\ell_j} \log\frac{V_j}{V_{j-1}};\end{aligned}\right\}\quad \begin{aligned}&0 \leqslant j \leqslant N,\\&m_j = 0\end{aligned} \tag{3.4.8}$$

$$\left.\begin{aligned}&|\varphi_j^{(\nu)}(\zeta)| \underset{F,\nu}{\lessapprox} V_j^{m_j-\nu},\ \nu \geqslant 0,\\&|\varphi_j^{(\nu)}(\zeta)| \underset{F,\nu}{\lessapprox} \sigma_j V_j^{m_j-\nu},\ m \geqslant m_j+1\end{aligned}\right\}\quad \begin{aligned}&0 \leqslant j \leqslant N,\\&m_j > 0.\end{aligned} \tag{3.4.9}$$

$$|\varphi_{N+1}^{(\nu)}(\zeta)| \underset{\nu,F}{\lessapprox} |\varphi_{N+1}(\zeta)|\, V_N^{-\nu}, \qquad \nu \geqslant 1,\quad \tilde{A} \geqslant A_{(3.4.10)}(F,n). \tag{3.4.10}$$

We differentiate (3.4.7) and obtain

$$f^{(n+1)}(\lambda) = \sum_{\substack{\nu_0+\dots+\nu_{N+1}=n+1\\ \nu_j \geqslant 0}} \frac{(n+1)!}{\nu_0!\dots\nu_{N+1}!}\, \varphi_0^{(\nu_0)}(\lambda)\dots\varphi_{N+1}^{(\nu_{N+1})}(\lambda). \tag{3.4.11}$$

Because $N \leqslant 2n+1$ we have only to estimate each term separately:

$$t_{\nu_0,\dots,\nu_{N+1}} \overset{\mathrm{def}}{=\!=} \varphi_0^{(\nu_0)}(\zeta)\dots\varphi_{N+1}^{(\nu_{N+1})}(\zeta).$$

Since we have

$$\sum_{j=0}^{N+1} \nu_j = n+1 > n \geqslant \sum_{j=0}^{N+1} m_j,$$

then there exists $j,\ 0 \leqslant j \leqslant N+1$ such that $\nu_j > m_j$. Let j_0 be the least such index. We cpnsider different possibilities using for estimating of the factor $\varphi_{\xi_0}^{(\nu_{j_0})}(\xi)$ the sharp inequalities from (3.4.8) and (3.4.9) and for estimating the rest factors rough inequalities.

1. $j = N+1$. Then $\nu_{N+1} > 0$ and

$$t_{\nu_0,\dots,\nu_{N+1}} \preccurlyeq \prod_{j=0}^{N} V_j^{m_j-\nu_j} \cdot Q\, V_N^{-\nu_{N+1}} \leqslant Q \prod_{j=1}^{N} V_N^{m_j-\nu_j} \cdot V_N^{-\nu_{N+1}} \asymp$$

$$\asymp \max_{D(x,V_N)} |P| \cdot V_N^{-n-1} \asymp \frac{V_N^n \log \frac{V_N}{y}}{V_N^{n+1}} \preccurlyeq \frac{1}{y}, \qquad Q \overset{def}{=} |\varphi_{N+1}(\zeta)|. \tag{3.4.12}$$

2. $j_0 \leqslant N,\ m_j > 0$. Then

$$t_{\nu_0,\dots,\nu_{N+1}} \preccurlyeq \prod_{0\leqslant j<j_0} V_j^{m_j-\nu_j} \cdot \sigma_{j_0} V_{j_0}^{m_{j_0}-\nu_{j_0}} \cdot \prod_{j_0<j\leqslant N} V_{j-1}^{m_j-\nu_j} \cdot Q\, V_N^{-\nu_{N+1}} \preccurlyeq$$

$$\preccurlyeq \prod_{0\leqslant j_0 \leqslant j} V_j^{m_j-\nu_j} \cdot \sigma_{j_0} V_{j_0}^{m_{j_0}-\nu_{j_0}} \cdot \prod_{j_0<j\leqslant N} V_j^{m_j} \cdot \prod_{j_0<j\leqslant N+1} V_{j_0}^{-\nu_j} \cdot Q \asymp$$

$$\asymp \sigma_j V_{j_0}^{-n-1} \max_{D(x,V_{j_0})} |P| \asymp \frac{\log(V_{j_0}/y)}{V_{j_0}} \preccurlyeq \frac{1}{y}. \tag{3.4.13}$$

3. $j_0 \leqslant N,\ m_j = 0$. Let

$$\tilde{\sigma}_j = \frac{V_j^n \log^2 \frac{4V_j}{y}}{\max\limits_{D(x,V_j)} |P|}, \qquad -1 \leqslant j \leqslant N.$$

Due to (3.4.8) we can write down the following estimate:

$$|\varphi_{j_0}^{(\nu_{j_0})}(\zeta)| \preccurlyeq \tilde{\sigma}_{j_0-1} V_{j_0-1}^{-\nu_{j_0}} + \tilde{\sigma}_{j_0} V_{j_0}^{-\nu_{j_0}}, \quad \nu_{j_0} \geqslant 1.$$

Hence

$$t_{\nu_0,\dots,\nu_{N+1}} \preccurlyeq \sum_{\delta=0}^{1} \prod_{0\leqslant j<j_0} V_j^{m_j-\nu_j} \cdot \tilde{\sigma}_{j_0-\delta} \cdot \prod_{j_0<j\leqslant N} V_{j-1}^{m_j-\nu_j} \cdot Q\, V_N^{-\nu_{N+1}} \preccurlyeq$$

$$\preccurlyeq \sum_{\delta=0}^{1} \prod_{0\leqslant j<j_0} V_{j_0-\delta}^{m_j-\nu_j} \cdot \tilde{\sigma}_{j_0-\delta} \cdot \prod_{j_0<j\leqslant N} V_j^{m_j} \cdot \prod_{j_0<j\leqslant N+1} V_{j_0-\delta}^{-\nu_j} \cdot Q \asymp$$

$$\asymp \sum_{\delta=0}^{1} \tilde{\sigma}_{j_0-\delta} V_{j_0-\delta}^{-n-1} \max_{D(x,V_{j_0-\delta})} |P| \asymp \sum_{\delta=0}^{1} \frac{\log^2 \frac{4V_{j_0-\delta}}{y}}{V_{j_0-\delta}} \preccurlyeq \frac{1}{y}. \tag{3.4.14}$$

The estimates (3.4.12) - (3.4.14) finish the proof of the part b) of Theorem 7. ●

§ 4. The embedding theorems of V.P.Havin- F.A.Shamoyan type

In this section we apply the results attained in the preceeding sections. Roughly speaking we intend to study the rate of the smoothness of an outer function f which corresponds to a function f with the summable logarithm on $\partial\mathbb{D}$ and belonging to an appropriate class of smoothness X. For some classes X it is possible to give a complete answer to that question for the functions $_e f$ and $_e f^2$. We recall that the first results of similar type were obtained by V.P.Havin and F.A.Shamoyan [12] and by V.P.Havin [13] .

THEOREM 8. Assume that f satisfies the condition

$$\int_{\partial\mathbb{D}} \log|f(\zeta)||d\zeta| > -\infty. \tag{A}$$

Then

$$f(\zeta)\in C^{\alpha}(\partial\mathbb{D}) \Rightarrow {}_e f\in \Lambda^{\alpha/2},\quad \alpha\neq 2n, \tag{3}$$

$$\not\Rightarrow {}_e f\in \Lambda^{\alpha/2+\varepsilon}, \tag{3'}$$

$$f(\zeta)\in L^p_{2n}(\partial\mathbb{D}) \Rightarrow {}_e f\in H^p_n,\ p>1,\ n\geq 1, \tag{4}$$

$$\not\Rightarrow {}_e f\in H^{p+\varepsilon}_n, \tag{4'}$$

$$f(\zeta)\in C^{\alpha}(\partial\mathbb{D}) \Rightarrow {}_e f^2\in \Lambda^{\alpha}, \qquad \alpha \text{ isn't integer} \tag{5}$$

$$\not\Rightarrow {}_e f^2\in \Lambda^{\alpha+\varepsilon}, \tag{5'}$$

$$f(\zeta)\in C^n(\partial\mathbb{D}) \Rightarrow {}_e f^2 \in \Lambda^{n-1} Z, \tag{6}$$

$$\not\Rightarrow {}_e f^2 \in \Lambda^n, \tag{6'}$$

$$f(\zeta)\in L^p_n(\partial\mathbb{D}) \Rightarrow {}_e f^2 \in H^{p/2}_n, \ p>2,\ n\geq 1, \tag{7}$$

$$\not\Rightarrow {}_e f^2 \in H^{p/2+\varepsilon}_n . \tag{7'}$$

REMARK. The implication (3) holds for $\alpha = 2n$ too; it $f\in L^p_{2n+1}$, $p>2$, then

$$\Big(\int_{\partial\mathbb{D}} |({}_e f)^{(n)}(e^{i(\theta+\delta)}) - ({}_e f)^{(n)}(e^{i\theta})|^p d\theta\Big)^{1/p} \leq C\delta^{1/2}.$$

The proof of these facts requires more techniques than the proof of the statements (3) - (7').

PROOF OF THEOREM (. By Theorems 5 - 7 it is sufficient to verify the following conditions:

The condition b) of Theorem 5 for the function f and for the number $\alpha/2$;

the condition b) of Theorem 6 for some function $\varphi\in L^p$ and for the number n;

the condition b) of Theorem 5 for the function f^2 and for the number α;

the condition b) of Theorem 7 for the function f^2 and for the number n;

the condition b) of Theorem 6 for some function $\psi\in L^{p/2}$ and for the number n.

Since

$$M_f^2(z) = M_{f^2}(z),$$

we have

$$M_f(z) \geq \rho^{\alpha/2} \Leftrightarrow M_{f^2}(z) \geq \rho^{\alpha},$$

$$M_f(z) \geq \varphi(z_0)\rho^{m/2} \Leftrightarrow M_{f^2}(z) \geq \varphi^2(z_0)\rho^m.$$

Hence it is sufficient to check only the condition b) of Theorem 5 for the function f and for the number $\alpha/2$ and the condition b) of Theorem 6 for some function $\varphi \in L^p$ and for the number $n/2$.

4.1. The verification of the condition b) of Theorem 5.

LEMMA 4.1. Let I be an arc, $I \subset \partial\mathbb{D}$, $f \in C^\alpha(I)$, $n < \alpha \leq n+1$, and moreover let

$$|f^{(n)}(\zeta) - f^{(n)}(z)| \leq |\zeta - z|^{\alpha - n}, \quad \zeta, z \in I.$$

Then there exists a number $A_\alpha > 0$ such that if

$$M \overset{\text{def}}{=} M_f(I) \geq A_\alpha |I|^\alpha$$

then

$$\int_I \log \frac{M}{|f(\zeta)|} |d\zeta| \underset{n}{\leq} |I|. \tag{4.1.1}$$

PROOF. The case $n = 0$ is trivial, we consider only the case $n > 1$. Let

$$\mu(h) = mes\{x \in I : |f(x)| \leq h\}$$

then

$$\int_I \log|f| |d\zeta| = |I| \log M - \int_0^M \frac{\mu(h)}{h} dh. \tag{4.1.1'}$$

Hence (4.1.1) is equivalent to the estimate

$$\int_0^M \frac{\mu(h)}{h} dh \underset{n}{\leq} |I|. \tag{4.1.2}$$

Let

$$E(h)=\{x\in I: |f(x)|\leqslant h\}, \quad 0\leqslant h\leqslant M.$$

There exist such points $x_0\leqslant\ldots\leqslant x_n$ that $x_j\in E(h)$, $0\leqslant j\leqslant n$, and

$$x_j - x_{j-1}\geqslant\frac{\mu(h)}{n}, \quad n>0.$$

Let P be the interpolating polynomial of f for the points

$$P(\zeta)=\sum_{j=0}^{n} f(x_j)\ \frac{(\zeta-x_0)\ldots\widehat{(\zeta-x_j)}\ldots(\zeta-x_n)}{(x_j-x_0)\ldots(x_j-x_n)}.$$

If $\zeta\in I$ then it is clear that

$$|P(\zeta)|\leqslant C_n\left(\frac{|I|}{\mu(h)}\right)^n h. \tag{4.1.3}$$

Next the function $\varphi=f-P$ has at least $n+1$ zeros on the arc I,

$$|\varphi^{(n)}(\zeta)-\varphi^{(n)}(z)|=|f^{(n)}(\zeta)-f^{(n)}(z)|\leqslant|\zeta-z|^{\alpha-n}.$$

Hence Lemma 1 from [11] implies

$$|\varphi(\zeta)|\leqslant C_\alpha |I|^\alpha.$$

Hence for $A_\alpha=2C_\alpha$ we obtain $M_P\geqslant\frac{1}{2}M$ and then (4.1.3) yields (4.1.2) follows:

$$\frac{1}{2}M\leqslant C_n\left(\frac{|I|}{\mu(h)}\right)^n h,$$

$$\mu(h)\underset{n}{\lesssim}|I|\left(\frac{h}{M}\right)^{1/n},$$

$$\int_0^M\frac{\mu(h)}{h}\,dh\underset{n}{\lesssim}|I|\int_0^M\frac{h^{\frac{1}{n}-1}}{M^{1/n}}\,dh\asymp|I|. \quad \bullet$$

We now proceed to the verification of the condition b) of Theorem 5 for $f \in C^{\alpha}(\partial\mathbb{D})$ and for the number $\alpha/2$. Without loss of generality we may assume that $|f(\zeta)| \leq 1$ and that

$$|f^{(n)}(\zeta) - f^{(n)}(z)| \leq |\zeta - z|^{\alpha - n}, \qquad \zeta, z \in \partial\mathbb{D}.$$

Let us take A_α from Lemma 4.1 and put

$$\sigma = \frac{1}{(2\pi A_\alpha)^{1/\alpha}} .$$

We suppose from now on that

$$M = M_f(z) \geq \rho^{\alpha/2}.$$

We have

$$\int_{\partial\mathbb{D}} |\log| \frac{M_f(z)}{f(\zeta)}|| \frac{1-|z|^2}{|\zeta - z|^2} |d\zeta| = \int_{|\zeta - z_0| > \sigma\sqrt{\rho}} + \int_{|\zeta - z_0| \leq \sigma\sqrt{\rho}} \overset{\text{def}}{=\!=} \mathcal{J}_1 + \mathcal{J}_2 .$$

Let $P = P_n(f; z_0; \cdot)$. Lemma 1.2 and the condition (A) yield

$$\mathcal{J}_1 = \int_{\substack{|\zeta - z_0| > \sigma\sqrt{\rho} \\ |f(\zeta)| \leq M}} + \int_{\substack{|\zeta - z_0| > \sigma\sqrt{\rho} \\ |f(\zeta)| > M}} \leq \frac{c}{\sigma^2} \int_{\partial\mathbb{D}} \log \frac{1}{|f|} |d\zeta| +$$

$$+ \int_{\partial\mathbb{D}} \log \Big[C_\alpha \frac{M_P(z) + M_P\big((1 - \frac{|\zeta - z|}{4}) z_0\big) + \rho^\alpha + |\zeta - z|^\alpha}{M} \Big] \frac{1-|z|^2}{|\zeta - z|^2} |d\zeta| \preccurlyeq 1 \qquad (4.1.4)$$

because we have due to $M \geq \rho^{\alpha/2} \geq \rho^\alpha$ that

$$M = M_f(z) \preccurlyeq M_P(z) + \rho^\alpha \preccurlyeq (M_f(z) + \rho^\alpha) + \rho^\alpha \preccurlyeq M .$$

By Lemma 1.2 and Lemma 4.1 (Lemma 4.1 is valid in view of the proper choice of σ) we obtain

$$\mathcal{J}_2 \preccurlyeq \sum_{\substack{n \geq 0 \\ 2^{n-1}\rho \leq \sigma\sqrt{\rho}}} 2^{-2n} \rho^{-1} \int_{|\zeta - z_0| \leq 2^n \rho} |\log| \frac{M}{f(\zeta)}|| |d\zeta| \leq$$

$$\leq \sum_{\substack{n \geq 0 \\ 2^{n-1}\rho \leq \sigma\sqrt{\rho}}} 2^{-2n} \rho^{-1} \int_{|\zeta - z_0| \leq 2^n \rho} |\log| \frac{M_f((1 - 2^n\rho) z_0)}{f(\zeta)}|| |d\zeta| +$$

$$\sum_{\substack{n\geq 0\\ 2^{n-1}\rho\leq\sigma\sqrt{\rho}}} 2^{-2n}\rho^{-1}\int\limits_{|\zeta-z|\leq 2^n\rho}\log\frac{M_f((1-2^n\rho)z_0)}{M}|d\zeta|\leq$$

$$\leq\sum_{n\geq 0}2^{-2n}\rho^{-1}\cdot 2^n\rho+\int\limits_{\partial\mathbb{D}}\log\left[C_\alpha\frac{M_f(z)+M_f((1-\frac{|\zeta-z|}{4})z_0)+\rho^\alpha+|\zeta-z_0|^\alpha}{M}\right]\cdot\frac{1-|z|^2}{|\zeta-z|^2}|d\zeta|\leq 1.$$

(4.1.5)

We see that (4.1.4) and (4.1.5) give the condition b) of Theorem 5 with the number $\alpha/2$. ●

4.2. The verification of the condition b) of Theorem 6.

LEMMA 4.2. Let I be an arc, $I\subset\partial\mathbb{D}$, $f\in L^p_n(I)$, $p>1$, $n\geq 1$,

$$x_1,\dots,x_n\in I,\quad f(x_j)=0,\qquad j=1,\dots,n,$$

$$F_\nu(x)=\frac{f(x)}{(x-x_1)\dots(x-x_\nu)},\qquad 1\leq\nu\leq n.$$

Then we have

$$\int\limits_I|F_\nu^{(n-\nu)}(\zeta)|^p|d\zeta|\underset{p,n}{\leq}\int\limits_I|f^{(n)}(\zeta)|^p|d\zeta|. \tag{4.2.1}$$

PROOF. First we state (4.2.1) for $\nu=1$ and for any $n\geq 1$. Let $P(x)=P_{n-1}(f;x_1;x)$, $\deg P\leq n-1$. Then

$$(P(t)/(t-x_1))^{(n-1)}\equiv 0$$

and

$$F_1^{(n-1)}(t)=\frac{1}{(n-1)!}\left[\frac{1}{t-x_1}\int\limits_{x_1}^{t}(t-y)^{n-1}f^{(n)}(y)\,dy\right]^{(n-1)},$$

(4.2.2)

and hence (4.2.2). Leibnitz formula and Hardy theorem yield (4.2.1) for the function F_1 . We now use the induction in n . For $n=1$

the statement (4.2.1) in just proved. Assume that it holds for $n=1$ and consider

$$f\in L^p_{n+1}(I), \quad f(x_1)=\ldots=f(x_{n+1})=0.$$

Let $f_1(x)=f(x)/(x-x_1)$. Then $f_1\in L^p_n(I)$, $f_1(x_2)=\ldots=f(x_{n+1})=0$.
If we put

$$F_{1\nu}(x)=f_1(x)/(x-x_2)\ldots(x-x_{\nu+1}), \quad 1\leqslant\nu\leqslant n,$$

then $F_\nu=F_{1,\nu-1}$, $\nu\geqslant 2$, and the preceeding estimates and the inductive assumption yield

$$\int\limits_I |F_\nu^{(n+1-\nu)}|^p=\int\limits_I |F_{1,\nu-1}^{(n-(\nu-1))}|^p \underset{n,p}{\lessapprox} \int\limits_I |f_1^{(n)}|^p \underset{n,p}{\lessapprox} \int\limits_I |f^{(n+1)}|^p. \quad \bullet$$

LEMMA 4.3. Let I be an arc,

$$I\subset\partial\mathbb{D}, \quad f\in L^q_n(I), \quad q>1, \quad n\geqslant 1$$

let the points

$$x_0, x_1,\ldots,x_n\in I, \quad f(x_\nu)=0, \quad 1\leqslant\nu\leqslant n, \quad \varphi=(f^{(n)})^*_q.$$

Then

$$|f(x)| \underset{q,n}{\lessapprox} |I|^n\varphi(x_0), \quad x\in I. \tag{4.2.3}$$

PROOF. If $n=1$, then

$$|f(x)|=\Big|\int\limits_{x_1}^{x} f'\Big|\leqslant |I|^{1/q'}\Big(\int\limits_I |f'|^q\Big)^{1/q}\leqslant |I|\varphi(x_0).$$

For $n>1$ we put

$$F(x)=\frac{f(x)}{(x-x_1)\ldots(x-x_{n-1})}.$$

Then Lemma 4.2 implies $F\in L^q_1(I)$, $F(x_n)=0$ and

$$|F(x)| \leq |I|^{1/r'} \Big(\int_I |F'|^r \Big)^{1/r} \underset{n,r}{\lesssim} |I|^{1/r'} \Big(\int_I |f^{(n)}|^r \Big)^{1/r} \leq |I| \varphi(x_0),$$

Hence

$$|f(x)| \underset{n,r}{\lesssim} |I|^n \varphi(x_0). \quad \bullet$$

LEMMA 4.4. Let I be an arc, $I \subset \partial\mathbb{D}$, $f \in L^r_n(I)$, $r>1$, $n \geq 1$, x_0 be the middle of $\varphi = (f^{(n)})^*_r$, $M = \max |I|$.

Then there exists $A = A(r,n) \geq 1$ such that if

$$M \geq A |I|^n \varphi(x_0)$$

then

$$\int_I \log \frac{M}{|f(\zeta)|} |d\zeta| \underset{n}{\lesssim} |I|. \tag{4.2.4}$$

PROOF. We denote

$$\mu(h) = \operatorname{mes} E_h, \quad E_h = \{x \in I : |f(x)| \leq h\}.$$

Using the formula (4.1.1') we conclude that it is sufficient to prove that

$$\int_0^M \frac{\mu(h)}{h} dh \underset{n}{\lesssim} |I|. \tag{4.2.5}$$

We take $C_{r,n}$ from (4.2.3) and put $A = A(r,n) = 2C_{r,n}$. We estimate $\mu(h)$, $0 < h \leq M$. We put $H = \mu(h)$. There exist n points $x_j \in E(h)$ such that $x_j - x_{j-1} \geq \frac{H}{n}$. Let P be the interpolating polynomial of the function f for the points $x_1, \dots, x_n$, $\varphi = f - P$. Lemma 4.2 and the choice of the number A yield

$$|\varphi(x)| \leq C_{r,n} |I|^n \varphi(x_0) \leq \frac{1}{2} M.$$

Hence if we have $|f(y)|=M$ at a point $y\in I$ then $|P(y)|\geqslant \frac{1}{2}M$. If $n=1$ then for $h\leqslant \frac{1}{4}M$ we would have

$$M=|f(y)|\leqslant|f(x_1)|+|\varphi(y)|\leqslant\frac{1}{4}M+\frac{1}{2}M=\frac{3}{4}M$$

because $P(y)\equiv f(x_1)$ in this case. Hence for $n=1$ we have $|f(x)|>\frac{1}{4}M,\ x\in I$, so (4.2.4) holds. For $n>1$ we have

$$\frac{1}{2}M\leqslant|P(y)|=\left|\sum_{j=1}^{n} f(x_j)\,\frac{(y-x_1)\dots\widehat{(y-x_j)}\dots(y-x_n)}{(x_j-x_1)\dots(x_j-x_n)}\right|\leqslant C_n h\left(\frac{|I|}{H}\right)^{n-1},$$

i.e.

$$\mu(h)=H\leqslant C_{n-1}|I|\left(\frac{h}{M}\right)^{\frac{1}{n-1}}$$

and that yields (4.2.5). ●

We are going now to verify the condition b) of Theorem 6 for a function $\varphi\in L^p(\partial\mathbb{D})$ chosen properly and for the number $n/2$. Let $f\in L^p_n(\partial\mathbb{D})$. We put $\varphi=(f^{(n)})^*_r$, where $1<r<p$. Then $\varphi\in L^p(\partial\mathbb{D})$. Suppose that

$$M=M_f(z)\geqslant\varphi\rho^{n/2}$$

where $\varphi=\varphi(z_0)$ and check that the condition (1) holds. Without loss of generality we assume $|f(\zeta)|\leqslant 1,\ \zeta\in\partial\mathbb{D}$. Let $P(\zeta)=$ $=P_{n-1}(f;z_0;\zeta)$. For a number $\sigma=\sigma(r,n)>0$, which we define later we can write by Lemma 1.2 that

$$\int_{\partial\mathbb{D}}=\int_{\partial\mathbb{D}\setminus D(z_0,\sigma\sqrt{\rho})}+\int_{\partial\mathbb{D}\cap D(z_0,\sigma\sqrt{\rho})}\stackrel{\text{def}}{=\!=}\mathcal{J}_1+\mathcal{J}_2\,;$$

$$\mathcal{J}_1=\int\limits_{\substack{|\zeta-z_0|>\sigma\sqrt{\rho}\\|f(\zeta)|\leqslant M}}\log\frac{M}{|f(\zeta)|}\,\frac{1-|z|^2}{|\zeta-z|^2}\,|d\zeta|+\int\limits_{\substack{|\zeta-z_0|>\sigma\sqrt{\rho}\\|f(\zeta)|>M}}\log\frac{|f(\zeta)|}{M}\cdot\frac{1-|z|^2}{|\zeta-z|^2}\,|d\zeta|\leqslant$$

$$\leqslant \frac{c}{\sigma^2}\int\limits_{\partial\mathbb{D}} \log\frac{1}{|f|}\,|d\zeta| + c\rho \int\limits_{|\zeta-z_0|>\sigma\sqrt{\rho}} \log\frac{M+|P(\zeta)|+C_{\tau,n}|\zeta-z_0|\varphi}{M}\,\frac{|d\zeta|}{|\zeta-z|^2} \preccurlyeq$$

$$\preccurlyeq \frac{1}{\sigma^2} + \sqrt{\rho}\,\log\left(1-\frac{\max\limits_{\Delta(z_0,\sigma\sqrt{\rho})}|P|}{M}\right) \preccurlyeq 1+\sqrt{\rho}\,\log\left(1+\rho^{-\frac{n-1}{2}}\right) \preccurlyeq 1. \tag{4.2.6}$$

In the last line of (4.2.6) we used the fact that

$$\max_{D(z_0,\sigma\sqrt{\rho})}|P| \underset{n,\sigma}{\leqslant} \frac{1}{\rho^{\frac{n-1}{2}}}\max_{D(z_0,2\rho)}|P| \underset{\sigma,n}{\preccurlyeq} \frac{M}{\rho^{\frac{n-1}{2}}}\,.$$

Let A be the number from Lemma 4.4. We put

$$\sigma = \left(\frac{1}{2\pi A}\right)^{2/n}\cdot\frac{1}{2\pi}\,.$$

For such a choice of σ we can apply Lemma 4.4 to any arc $\gamma = \partial\mathbb{D}\cap D(z_0,\delta)$, $\rho\leqslant\delta\leqslant\sigma\sqrt{\rho}$. Since

$$M_f(\gamma) \geqslant M_f(z) \geqslant A|\gamma|^n\varphi(z_0).$$

Hence

$$\mathcal{I}_2 = \int\limits_{|\zeta-z_0|\leqslant\rho} + \sum_{\substack{N\geqslant 0\\ 2^{N+1}\rho\leqslant\sigma\sqrt{\rho}}} \int\limits_{2^N\rho\leqslant|\zeta-z_0|\leqslant 2^{N+1}} + \int\limits_{\frac{\sigma}{2}\sqrt{\rho}\leqslant|\zeta-z_0|\leqslant\sigma\sqrt{\rho}}\,. \tag{4.2.7}$$

Let $M_N = \max\limits_{\partial\mathbb{D}\cap D(z_0,2^N\rho)}|f|$. Then

$$M_N \leqslant \max_{D(z_0,2^N\rho)}|P| + C_n\cdot(2^N\rho)^n\varphi(z_0) \preccurlyeq 2^{Nn}\max_{D(z_0,\rho)}|P| + \rho^{\frac{n}{2}}\varphi(z_0) \preccurlyeq 2^{Nn}M,$$

Hence by Lemma 4.4 for $2^{N+1}\rho\leqslant\sigma\sqrt{\rho}$ we have

$$\int\limits_{2^N\rho\le|\zeta-z_0|\le 2^{N+1}\rho} \log\left|\frac{M}{f(\zeta)}\right|\left|\frac{1-|z|}{|\zeta-z|^2}\right||d\zeta|\le\frac{1}{2^{2N}\rho}\int\limits_{|\zeta-z|\le 2^{N+1}\rho}\left|\log\left|\frac{M}{f(\zeta)}\right|\right||d\zeta|\le$$

$$\le\frac{1}{2^{2N}\rho}\left(\int\limits_{|\zeta-z_0|\le 2^{N+1}\rho}\log\frac{M_{N+1}}{|f(\zeta)|}|d\zeta|+2^N\rho\log\frac{M_{N+1}}{M}\right)\le\frac{1}{2^{2N}\rho}(2^N\rho+2^N\rho\cdot N).$$

Hence

$$\sum_{\substack{N\ge 0\\ 2^{N+1}\rho\le 6\sqrt{\rho}}}\le\sum_{N=0}^{\infty}\frac{N+1}{2^N}\le 1. \tag{4.2.8}$$

Similarly we obtain

$$\int\limits_{|\zeta-z_0|<\rho}+\int\limits_{\frac{6}{2}\sqrt{\rho}\le|\zeta-z_0|\le 6\sqrt{\rho}}\le 1. \tag{4.2.9}$$

We see that (4.2.6) - (4.2.9) yield the estimate (1) by the assumption

$$M=M_f(z)\ge(f^{(n)})^*_{\eta}(z_0)\cdot\rho^{\frac{n}{2}},$$

and that finish the proof of (4) and (7) of Theorem 8.

4.3. The strictness of the inclusions in Theorem 8.

In this section we present the main features of the proofs of (3') - (7') which by means of relatively standard but slightly long arguments could easily be reduced to the correct form (5') - and (6'): there exists a function $h\ge 0$ such that $h\in C^d_h$ or $h\in C^n$ but $\tilde{h}\notin C^{d+\varepsilon}$ or $\tilde{h}\notin C^n$. Let $f=e^h$. Then the function $e f^2=e^{2(h+i\tilde{h})}$ does not belong respectively to $\Lambda^{d+\varepsilon}$ or Λ^n.

(3'): to simplify the notation we work on the real line $\mathbb{R}$. Let $f\in C^{(d)}(\mathbb{R})$, $f\ge 0$ and

$$f(x)=\begin{cases} x^{d}, & x>0 \\ v(x), & x<0 \end{cases} \tag{4.3.1}$$

where v has a sufficiently regular behaviour and is small enough for the function

$$\mu(y)=\int_{-y}^{0} \log v(x)\,dx$$

tends to zero as $y \to 0$ as slowly as we want. Then after some computations we obtain

$$({}_e f)^{(N)}(x+ix^2)=\lambda(x)x^{d-2N}=\lambda(x)(x^2)^{\frac{d}{2}-N}, \tag{4.3.2}$$

where $0<x<1$, $N>\frac{d}{2}$, $\lambda(x)$, tends to zero very slowly too (the speed of vanishing of λ depends on that of μ). So (4.3.2) shows that if μ is chosen in such a way that $\lambda(x) \asymp 1/\log\frac{2}{x}$, then $f \notin \Lambda^{d/2+\varepsilon}$ for an arbitrary $\varepsilon>0$.

In the following counterexamples we take a function v which like in (4.3.1) has a regular behaviour and tends to zero very rapidly.

(4'): We put

$$f(x)=\begin{cases} x^{2n-\frac{1}{p}}\log\frac{1}{x}, & x>0, \\ v(x), & x<0 \end{cases}$$

and demand $f \in L^{p}_{2n}(\mathbb{R})$. Then

$$({}_e f)^{(n)}(x)=\lambda(x)x^{-\frac{1}{p}}\log\frac{2}{x}, \quad 0<x<1, \quad \lambda(x)\downarrow 0$$

and $({}_e f)^{(n)} \notin H^{p+\varepsilon}$ if $\lambda(x) \asymp 1/\log\frac{2}{x}$

(7'): We put

$$f(x)=\begin{cases} x^{n-\frac{1}{p}}\log\frac{1}{x}, & x>0, \\ v(x), & x<0 \end{cases}$$

where $f \in L^p_n(\mathbb{R})$. Then

$$({}_e f^2)^{(n)}(x) = \lambda(x)\, x^{-\frac{2}{p}} \log\frac{2}{x}, \quad 0 < x < 1$$

and ${}_e f^2 \notin H_n^{p/2+\varepsilon}$ if $\lambda(x) \asymp 1/\log\frac{2}{x}$.

These counterexamples finish the proof of Theorem 8. ●

4.4. The corollary from Theorems 5 and 6.

COROLLARY. Let $X = \Lambda^\alpha$, $0<\alpha<1$ or $X = H_1^p$, $1 < p < \infty$. Suppose that $f_j \in X$ and $\sum \|f_j\|_X < \infty$ (*)

We put

$$f(\zeta) = \sum_j |f_j(\zeta)|, \quad \zeta \in \partial\mathbb{D}.$$

Then ${}_e f \in X$.

In particular if $f_1, f_2 \in X$ and $f(\zeta) = |f_1(\zeta)| + |f_2(\zeta)|$, $\zeta \in \partial\mathbb{D}$ then ${}_e f \in X$.

PROOF. We need two auxiliary statements which are easy consequences of Theorems 5 and 6 (see too [10]).

LEMMA 4.5. (a form of Theorem 5 for $0<\alpha<1$). Let $h \in C(\partial\mathbb{D})$, $h \geqslant 0$. Then ${}_e h \in \Lambda^\alpha$ if and only if $h \in C^\alpha(\partial\mathbb{D})$ and for any $z_0 \in \partial\mathbb{D}$ such that $h(z_0) \geqslant \delta\rho^\alpha$, $\delta > 0$,

$$\int_{\partial\mathbb{D}} \log^+ \frac{h(z_0)}{h(\zeta)} \cdot \frac{1-|z|^2}{|\zeta - z|^2}\,|d\zeta| \leqslant C(\alpha, \delta, \|h\|_{C^\alpha}), \quad z = (1-\rho) z_0. \tag{4.4.1}$$

LEMMA 4.6. (a form of Theorem 6 for H_1^p)

a) Let $f \in H_1^p$, $1<p<\infty$, $\psi = (f')^*$. There exists an absolute constant A such that if $|f(z_0)| \geqslant A\psi(z_0)\rho$, $z_0 \in \partial\mathbb{D}$ then

$$\int_{\partial\mathbb{D}} \log^+ \frac{h(z_0)}{h(\zeta)} \cdot \frac{1-|z|^2}{|\zeta-z|^2}\,|d\zeta| \leqslant C(p, \|f\|_{H^p}), \quad z = (1-\rho) z_0, \tag{4.4.2}$$

where $C(\rho, \|f\|_{H^p})$ depends only on ρ and $\|f\|_{H^p}$.

b) function $h \geq 0$, $h \in L^p_1(\partial\mathbb{D})$ such that for any point $z_0 \in \partial\mathbb{D}$ the condition $h(z_0) \geq \psi(z_0)$ implies (4.4.2) then ${}_e h \in H^p_1$.

PROOF OF THE COROLLARY.

1. $X = \Lambda^\alpha$. It is clear that (*) implies $f \in C^\alpha(\partial\mathbb{D})$. Let us check (4.4.1). Assume that $f(z_0) \geq \rho^\alpha$, $z_0 \in \partial\mathbb{D}$ and let the numbers $\theta_j = \theta_j(z_0)$, $|\theta_j| = 1$ be such that $\theta_j f_j(z_0) = |f_j(z_0)|$. We introduce an auxiliary function $\Phi(t) = \Phi_{z_0}(t) = \sum \theta_j f_j(t)$. Then (*) yields $\Phi \in \Lambda^\alpha$, $\|\Phi\|_{\Lambda^\alpha} \leq C$, C does not depend on the choice of z_0. But $\Phi(z_0) = f(z_0) \geq \rho^\alpha$ hence

$$\int_{\partial\mathbb{D}} \log^+ \left|\frac{\Phi(z_0)}{\Phi(\zeta)}\right| \frac{1-|z|^2}{|\zeta - z|^2} |d\zeta| \leq C_1, \quad z = (1-\rho) z_0, \tag{4.4.3}$$

where C_1 depends only on C and α. But

$$|\Phi(\zeta)| \leq \sum_j |f_j(\zeta)| = f(\zeta), \quad \zeta \in \partial\mathbb{D}.$$

Hence

$$\log^+ \frac{f(z_0)}{f(\zeta)} \leq \log^+ \left|\frac{\Phi(z_0)}{\Phi(\zeta)}\right|, \quad \zeta \in \partial\mathbb{D}. \tag{4.4.4}$$

Now (4.4.3) and (4.4.4) imply (4.4.1) for f, i.e. ${}_e f \in \Lambda^\alpha$.

2. $X = H^p_1$. We take the number A from Lemma 4.6 and put $\psi_j = A(f_j')^*$, $\psi = \sum \psi_j$. Then (*) and maximal theorems for L^p yield $\psi \in L^p(\partial\mathbb{D})$. Assume that $f(z_0) \geq \psi(z_0)\rho$, $z_0 \in \partial\mathbb{D}$. We choose the numbers $\theta_j = \theta_j(z_0)$, $|\theta_j| = 1$ such that $\theta_j f_j(z_0) = |f_j(z_0)|$, and introduce the function $\Phi(t) = \Phi_{z_0}(t) = \sum \theta_j f_j(t)$. Then (*) implies $\Phi \in H^p_1$. Furthermore $|\Phi'|^* \leq \sum (f_j')^*$, $\Phi(z_0) = f(z_0)$. Hence $f(z_0) \geq \psi(z_0)\rho$ yields $\Phi(z_0) \geq A(\Phi')^*(z_0)\rho$ and then due to (4.4.2)

$$\int_{\partial\mathbb{D}} \log^+ \left|\frac{\Phi(z_0)}{\Phi(\zeta)}\right| \frac{1-|z|^2}{|\zeta - z|^2} |d\zeta| \leq C, \quad z = (1-\rho) z_0, \tag{4.4.5}$$

where C depends only on p and $\|\Phi_{z_0}\|_{H_1^p}$, i.e. C does not depend on z_0. But $|\Phi(\zeta)| \leq f(\zeta)$ hence

$$\log^+ \frac{f(z_0)}{f(\zeta)} \leq \log^+ \left|\frac{\Phi(z_0)}{\Phi(\zeta)}\right| \tag{4.4.6}$$

and (4.4.5), (4.4.6) and (4.4.2) give $e^f \in H_1^p$. ●

CHAPTER III

ZEROS AND THEIR MULTIPLICITIES

In this chapter we deal with the zero-sets of analytic functions as well as with their behaviour near zero-sets. In § 1 we describe the zero-sets of functions from the class Λ_ω , ω being an arbitrary modulus of continuity, and obtain a natural generalization of the well known Beurling-Carleson condition in that case. In § 2 we study possible multiplicities of zeros on the unit circle for the functions with either rare or fast decreasing Taylor-coefficient and find ultimate estimates.

§ 1. Zero-sets of functions from Λ_ω

Let $f \in \Lambda_\omega$, $f = {}_e f \cdot I_f$ be the Nevanlinna-factorization of f , $E = Z_f(\overline{\mathbb{D}})$. It is clear that

$$\operatorname{spec} I_f \subset E, \tag{1.0}$$

$$\sum_{\alpha \in E} (1-|\alpha|^2) < \infty. \tag{1.1}$$

Applying the Jensen formula we obtain that we also have

$$\int_{\partial\mathbb{D}} \log \omega(\operatorname{dist}(z,E))\,|dz| > -\infty. \tag{1.2a}$$

We show that the estimates (1.0) - (1.2a) are sufficient for the existence of a function $f \in \Lambda_\omega$, $f \not\equiv 0$ with the zero-set E for any modulus of continuity ω with possible non-regular behaviour (in case of ω with regular behaviour, more explicitly, in the case when ω is a finite iterate of the logarithms it was proved by J. Stegbuchner [17] , [18]).

THEOREM 9. Let S, $S \in \mathcal{J}_b$ be an inner function, $\operatorname{spec} S \subset E$, where E satisfies (1.1) and (1.2$_\omega$), ω be an arbitrary modulus of continuity. Then there exists a function $f \in \Lambda_\omega$ such that $Z_f(\bar{\mathbb{D}}) = E$ and $f/S \in \Lambda_\omega$.

REMARK. Using slightly more complicated techniques we can construct a function $f \in \Lambda_\omega$ such that

for which for $N = 1, 2, \ldots$ we have

$$|f(z)| = O(\omega^N(\operatorname{dist}(z, E \cap \partial\mathbb{D}))), \quad z \in \bar{\mathbb{D}},$$

where the constant in the $\operatorname{sign} O$ depends on N .

1.1. The reduction to the main lemma.

We first describe a geometric construction which permits us to restrict our attention to the sets E with the following additional property:

$$\operatorname{dist}(z, E \cap \partial\mathbb{D}) \asymp 1 - |z|, \quad z \in E \cap \mathbb{D}. \tag{1.1.1}$$

In case the set E does not satisfy (1.1.1) but possesses the properties (1.1) and (1.2$_\omega$) we act as follows (see [16]). Let $E_0 = E \cap \partial\mathbb{D}$, $E_1 = E \cap \mathbb{D}$ and enlarge the set E_0 by a countable set in a such a way that we a new set $E_0^* \supset E_0$, $E_0^* \subset \partial\mathbb{D}$ obtained satisfies: the set $E^0 = E_0^* \setminus E_0$ consists of the isolated points and the set $E^* = E_0^* \cup E_1$ possesses the properties (1.1), (1.2$_\omega$) and (1.1.1). To any point $z^* \in E^0$ we put into correspondence the point $\zeta^* = R(z^*)z^*$, $R(z^*) > 1$ and the region G_{z^*} which is bounded by smaller arc of the circle Γ_{z^*} of radius $\frac{1}{2}$ touching the tangents coming from the point ζ^* to $\partial\mathbb{D}$, by the segments of these tangents between the touching-points of $\partial\mathbb{D}$ and Γ_{z^*} and by the bigger arc of $\partial\mathbb{D}$ between the touching points. We denote by $\tilde{\xi}$ the point of intersection of the segment $[z^*, \zeta^*]$ with ∂G_{z^*} and put $\tilde{E} = \bigcup_{z^* \in E_0^\circ} \{\tilde{\xi}\}$. We can choose the numbers $R(z^*)$ so close to 1 that the regions G_{z^*} are disjoint for different z^*. We put $G = \bigcup_{z^* \in E^\circ} G_{z^*}$. We map conformally the region G onto $\mathbb{D}$ by the function φ. For $R(z^*)-1$ sufficiently small the boundary of G is the Liapunov-curve. Hence we can apply the Kellog theorem to the map φ [33] and obtain that the set $H^* = \varphi(\tilde{E} \cup E)$ satisfies (1.1), (1.2$_\omega$), (1.1.1). Moreover, if $\sigma(\zeta)$ is the inner factor of $S(\varphi^{-1}(\zeta))$, S is taken from the assumptions of Theorem 9, then $\sigma(\zeta)/S(\varphi^{-1}(\zeta))$ is C^1-smooth in $\overline{\mathbb{D}}$ for the proper choice of $R(z^*)$. Assume that for the set H^* and for the function σ we have constructed a function $F \in \Lambda_\omega$ such that $Z_F(\overline{\mathbb{D}}) = H^*$ and $F/\sigma \in \Lambda_\omega$.

Let $f(z) = F(\varphi^{-1}(z))$. Then $f \in \Lambda_\omega$, $Z_f(\overline{\mathbb{D}}) = E$, $f/S \in \Lambda_\omega$, q.e.d.

So we can assume without loss of generality that the set E possesses the properties (1.1), (1.2$_\omega$), (1.1.1). Let $\{I_n\}_n$ be the union of all supplementary arcs of the set E_0, ζ_n being the middle of I_n.

MAIN LEMMA. There exist numbers $\lambda_n > 0$ with the following properties

1.

$$\lambda_n \leq \min\left(1, \omega\left(\frac{|I_n|}{2\pi}\right), \omega\left(\frac{1}{|S'(\zeta_n)|}\right)\right),$$

2.

$$\sum_n |I_n| \log \lambda_n > -\infty,$$

3.
$$c|I_n|^2 \leq \lambda_n \leq \omega\left(\frac{1}{a_n}\right),$$

where $c>0$ does not depend on n and

$$a_n = \int\limits_{\partial\mathbb{D}\setminus I_n} \frac{v(t)|dt|}{|t-\zeta_n|^2}$$

and the function $v(t)$ for $t=e^{i\theta}\in I_m = (e^{i\alpha}, e^{i\beta})$ is defined as follows:

$$v(t) = \log\frac{1}{\lambda_m} + \log\frac{(\beta-\alpha)^2}{(\beta-\theta)(\theta-\alpha)}.$$

We prove the main lemma in subsection 1.2. Assuming for a while the main lemma proved, we check now that the function $f=FS$ belongs to Λ_ω where F is an outer function which we define by means of its modulus for $e^{i\theta}\in I_n = (e^{i\alpha}, e^{i\beta})$ with the help of the following formula:

$$|F(e^{i\theta})| = \lambda_n \frac{(\theta-\alpha)^2(\beta-\theta)^2}{(\beta-\alpha)^4}.$$

Moreover, the construction yields $Z_f(\bar{\mathbb{D}}) = E$ and $F\in C_A$ and that implies due to Theorem 1 that $F = f/S \in \Lambda_\omega$, i.e. f is the required function.

The definition of F is correct because of the property 2, when verifying the fact that $F\in\Lambda_\omega$ we can consider by P.M.Tamrazov's theorem [28] only the points lying an $\partial\mathbb{D}$, i.e. it is sufficiently to check

$$|f(z_1)-f(z_2)| \leq c_f\,\omega(|z_1-z_2|), \quad z_1,z_2\in\partial\mathbb{D}.$$

If $z_1\in I_1$, $z_2\in I_2$ where I_1, I_2 are different complementary arcs of the set $E\cap\partial\mathbb{D}$ then the property 1, and the definition of $|F|$ gives

$$|f(z_1)-f(z_2)| \leqslant |f(z_1)|+|f(z_2)| \leqslant$$

$$\leqslant const \cdot \max_{j=1,2}\left(\omega\left(\frac{|I_j|}{2\pi}\right)\cdot \frac{dist\,(z_j,E_0)}{|I_j|}\right) \leqslant$$

$$\leqslant const \cdot \max_{j=1,2} \omega(dist\,(z_j,E_0)) \leqslant const \cdot \omega(|z_1-z_2|). \tag{1.1.2}$$

But if $z_1, z_2 \in I_n \subset \partial\mathbb{D} \setminus E_0$ and $\delta \overset{def}{=} |z_1-z_2| \geqslant \frac{1}{2} dist\,(z_1,z_2)$ then applying the property 1 again we have as in (1.1.2) that

$$|f(z_1)-f(z_2)| \leqslant |f(z_1)|+|f(z_2)| \leqslant const \cdot \omega(|z_1-z_2|). \tag{1.1.3}$$

Assume now that $\delta = |z_1-z_2| \leqslant \frac{1}{2} dist\,(z_1, E_0)$. We put $\rho(z) = = dist\,(z, E_0)$ and write

$$f(z_1)-f(z_2) = (F(z_1)-F(z_2))S(z_1)+F(z_2)(S(z_1)-S(z_2)) \overset{def}{=} T_1+T_2.$$

The estimate of T_1 . It $\delta > \frac{1}{a_n}\cdot\frac{\rho^2(z_1)}{|I_n|^2}$ then 1. yields

$$|T_1| \leqslant const \cdot \lambda_n \frac{\rho^2(z_1)}{|I_n|^2} \leqslant const \cdot \omega\left(\frac{1}{a_n}\right)\cdot \frac{\rho^2(z_1)}{|I_n|} \leqslant const \cdot \omega(\delta). \tag{1.1.4}$$

But if $\delta < \frac{1}{a_n} \frac{\rho^2(z_1)}{|I_n|^2}$ then the natural estimates of the Schwarz's integral lead us to the following estimate:

$$|T_1| = |F(z_1)|\cdot\left|1-\frac{F(z_2)}{F(z_1)}\right| \leqslant$$

$$\leq const \cdot \lambda_n \frac{\rho^2(z_1)}{|I_n|^2} \cdot \delta \cdot \left(\frac{|I_n|^2}{\rho^2(z_1)} a_n + \frac{1 + \log \frac{|I_n|}{\rho(z_1)}}{\rho(z_1)} + \frac{|\log \lambda_n|}{\rho(z_1)} \right). \tag{1.1.5}$$

The property 3. yields

$$\frac{|\log \lambda_n|}{\rho(z_1)} \leq const \frac{|\log |I_n||}{\rho(z_1)}$$

and the definition of a_n implies

$$a_n \geq const \cdot \frac{|\log |I_n||}{|I_n|}$$

hence the property 3 again implies

$$(1.1.5) \leq const \cdot (\delta \lambda_n a_n + \frac{\delta}{|I_n|} \omega(|I_n|)) \leq$$

$$\leq const \cdot \left(\delta \omega \left(\frac{1}{a_n} \right) a_n + \frac{\delta}{|I_n|} \omega(|I_n|) \right) \leq const \cdot \omega(\delta). \tag{1.1.6}$$

The estimate of the term T_2. If $\delta \geq \frac{1}{|S'(\zeta_n)|} \cdot \frac{\rho^2(z_1)}{|I_n|^2}$ then 1. gives

$$|T_2| \leq const \cdot \lambda_n \frac{\rho_2(z_1)}{|I_n|^2} \leq const \cdot \omega \left(\frac{1}{|S'(\zeta_n)|} \right) \cdot \frac{\rho^2(z_1)}{|I_n|^2} \leq const \cdot \omega(\delta). \tag{1.1.7}$$

But if $\delta < \frac{1}{|S'(\zeta_n)|} \cdot \frac{\rho_2(z_1)}{|I_n|}$ then 1. and (1.1.1) yield

$$|T_2| \leq const \cdot \lambda_n \frac{\rho^2(z_1)}{|I_n|^2} \cdot \delta |S'(z_1)| \leq const \cdot \delta \omega \left(\frac{1}{|S'(\zeta_n)|} \right) \cdot |S'(\zeta_n)| \leq const \cdot \omega(\delta). \tag{1.1.8}$$

The estimates (1.1.2) - (1.1.8) show that $f \in \Lambda_\omega$. ●

1.2. Proof of the main lemma.

We begin from an auxiliary fact.

LEMMA 1.1. Let $E = \overline{E} \subset \overline{\mathbb{D}}$, $E_0 = E \cap \partial \mathbb{D}$, $E_1 = E \cap \mathbb{D}$,

$dist\,(z_1, E_0) \asymp 1 - |z_1|$ for $z_1 \in E_1$, $S \in \mathcal{J}_b$ be an inner function, $spec\, S \subset E$ and

$$\int_{\partial\mathbb{D}} \log \omega\,(dist\,(z, E_0))\,|dz| > -\infty .$$

Then

$$\int_{\partial\mathbb{D}} \log \omega \left(\frac{1}{|S'(z)|} \right) |dz| > -\infty .$$

PROOF. Let $\{I_n\}$ be the complementary arcs of the set E_0, being the middle of I_n. Then

$$\int_{\partial\mathbb{D}} \log \omega \left(\frac{1}{|S'(z)|} \right) |dz| = \sum_n |I_n| \log \omega \left(\frac{1}{|S'(\zeta_n)|} \right) + O(1). \tag{1.2.1}$$

For the sake of brevity we denote by $A_1, A_2, \dots, C_1, C_2, \dots$ constants which do not depend on n, m or N. We have

$$\sum_n |I_n| \log \omega \left(\frac{1}{|S'(\zeta_n)|} \right) = \sum_{|I_n| \leqslant \frac{1}{|S'(\zeta_n)|}} + \sum_{|I_n| > \frac{1}{|S'(\zeta_n)|}} \geqslant$$

$$\geqslant \sum_{|I_n| \leqslant \frac{1}{|S'(\zeta_n)|}} |I_n| \log \omega (|I_n|) + \sum_{|I_n| > \frac{1}{|S'(\zeta_n)|}} |I_n| \log \omega (|I_n|) +$$

$$+ \sum_{|I_n| > \frac{1}{|S'(\zeta_n)|}} |I_n| \log \frac{\omega\left(\frac{1}{|S'(\zeta_n)|}\right)}{\omega(|I_n|)} \geqslant A_1 + \sum_{|I_n| > \frac{1}{|S'(\zeta_n)|}} |I_n| \log \frac{1}{|I_n||S'(\zeta_n)|} . \tag{1.2.2}$$

We introduce now an auxiliary pure singular function S_0, $S_0 \in \mathcal{J}_b$ such that $spec\, S_0 \subset E$ and $|S_0'(\zeta_n)| \asymp |S'(\zeta_n)|$, $n = 1, 2, \dots$. For that purpose to every zero b, $|b| < 1$ of the function S we can accord the pure singular factor with the spectrum contained at the point of E_0 nearest to b and with measure at that point equal to $1 - |b|$. Let

$$M_0(z) \overset{def}{=\!=} \sup_{z\in\gamma\subset\partial\mathbb{D}} \frac{1}{|\gamma|}\int_\gamma d\mu_0 + 1, \quad z\in\partial\mathbb{D}, \tag{1.2.3}$$

where $d\mu_0$ is the singular measure defining S_0 and sup in (1.2.3) is taken over all arcs $\gamma\ni z$. Then

$$|I_n|\log\frac{1}{|I_n||S_0'(\zeta_n)|} \geq \int_{I_n}\log\frac{1}{M_0(z)}|dz| - c|I_n|,$$

and hence the classic theorems [34] Ch.1 yields

$$(1.2.2) \geq A_2 - \int_{\partial\mathbb{D}}\log M_0(z)|dz| \geq A_2 - c_1\int_{\partial\mathbb{D}}\sqrt{M_0(z)}|dz| \geq A_2 - c_2\sqrt{\mu_0(\partial\mathbb{D})}. \qquad \bullet$$

We proceed now to the inductive construction of the sequence $\{\lambda_n\}$. Let

$$\mu_n = \min\left(\omega\left(\frac{|I_n|}{2\pi}\right), \quad \omega\left(1/|S'(\zeta_n)|\right)\right).$$

Then (1.2.1) and (1.2) imply

$$\sum_n |I_n|\log\mu_n > -\infty. \tag{1.2.4}$$

We put $\lambda_{n0}=\mu_n$ and define inductivly the sequences λ_{nN}, a_{nN} and the sequence of functions v_N as follows:

$$v_N(t) = \log\frac{1}{\lambda_{mN}} + \log\frac{(\beta-\alpha)^2}{(\beta-\theta)(\theta-\alpha)}, \quad t=e^{i\theta}\in I_m = (e^{i\alpha}, e^{i\beta}),$$

$$N=0,1,2,\dots;$$

$$a_{nN} = \int_{\partial\mathbb{D}\setminus I_n}\frac{v_N(t)|dt|}{|t-\zeta_n|^2}, \quad N=0,1,\dots; \tag{1.2.5}$$

$$\lambda_{n,N+1} = \min\left(\lambda_{nN},\ \omega\left(\frac{1}{a_{nN}}\right)\right). \tag{1.2.6}$$

The inductive definition (1.2.6) of the double-indexed sequence λ_{nN} shows that λ_{nN} decreases for any fixed n as N increases. Hence there exists a limit $\lambda_n = \lim_{N\to\infty} \lambda_{nN}$. We assert that $\{\lambda_n\}$ is the required sequence.

For the proof we define for every n number $M(n) \leqslant +\infty$ as follows:

If for all $N \geqslant 1$ we have $\omega(1/a_{n,N-1}) > \lambda_{n0}$ then we put $M(n) = +\infty$. If $M(n) > 1$ then for $1 \leqslant N \leqslant M(n)-1$ we have $\omega\left(\frac{1}{a_{n,N-1}}\right) > \lambda_{n0}$.

Hence $\lambda_{nN} = \lambda_{n0}$ for $1 \leqslant N \leqslant M(n)-1$. If $M(n) < \infty$ then $\lambda_{n,M(n)} = \omega\left(1/a_{n,M(n)-1}\right)$ and because $a_{n,M(n)} \geqslant a_{n,M(n)-1}$ we have $\omega(1/a_{n,M(n)}) \leqslant \leqslant \omega\left(1/a_{n,M(n)-1}\right)$.

Hence $\lambda_{n,M(n)+1} = \omega(1/a_{n,M(n)})$ and then we get inductively $\lambda_{nN} = \omega(1/a_{n,N-1})$, $N \geqslant M(n)$.

We get now from (1.2) and (1.2.4):

$$\begin{aligned}
\sum_n |I_n| \log\frac{1}{\lambda_{n,N}} &= \sum_{M(n)\leqslant N} + \sum_{M(n)>N} = \sum_{M(n)>N} |I_n| \log\frac{1}{\mu_n} + \\
&+ \sum_{M(n)\leqslant N} |I_n| \log\frac{1}{\omega(1/a_{n,N-1})} = \sum_{M(n)>N} |I_n| \log\frac{1}{\mu_n} + \sum_{M(n)\leqslant N} |I_n| \log\frac{1}{\omega\left(\frac{|I_n|}{2\pi}\right)} + \\
&+ \sum_{M(n)\leqslant N} |I_n| \log\frac{\omega\left(\frac{|I_n|}{2\pi}\right)}{\omega(1/a_{n,N-1})} \leqslant A_1 + \sum_{M(n)\leqslant N} |I_n| \log(|I_n|\cdot a_{n,N-1}) \leqslant \\
&\leqslant A_1 + \sum_n |I_n| \log(1 + |I_n|\cdot a_{n,N-1}).
\end{aligned} \tag{1.2.7}$$

Let

$$M_{N-1}(t) \stackrel{\text{def}}{=\!=} \sup_{t\in\gamma\subset\partial\mathbb{D}} \frac{1}{|\gamma|} \int_\gamma (v_{N-1}(z)+1)|dz|, \quad t\in\partial\mathbb{D}.$$

We notice that

$$|I_n| \log(1+|I_n| a_{n,N-1}) \leq \int_{I_n} \log M_{N-1}(t)|dt| + c|I_n| \tag{1.2.8}$$

and

$$\int_{\partial \mathbb{D}} (v_N(t)+1)|dt| \geq \sum_n |I_n| \log \frac{1}{\lambda_{nN}} \geq \int_{\partial \mathbb{D}} (v_N(t)+1)|dt| - A_2 . \tag{1.2.9}$$

Let $S_N = \int_{\partial \mathbb{D}} (v_N(t)+1)|dt|$. Then (1.2.7) - (1.2.9) and the classic theorem [34] , Ch. 1 yield

$$S_N \leq A_3 + \int_{\partial \mathbb{D}} \log M_{N-1}(t)|dt| \leq A_3 + c \int_{\partial \mathbb{D}} \sqrt{M_{N-1}(t)}\, |dt| \leq$$

$$\leq A_3 + c_1 \sqrt{\int_{\partial \mathbb{D}} (v_{N-1}(t)+1)|dt|} = A_3 + c_1 \sqrt{S_{N-1}} . \tag{1.2.10}$$

Because due to (1.2.4) $S_0 < \infty$ we obtain that (1.2.10) implies $S_N < \infty$ for all $N = 1, 2, \ldots$. Further $S_{N-1} \leq S_N$. Hence $S_N \leq A_3 + c_1 \sqrt{S_N}$ which yields $S_N \leq A_4$ and bearing in mind that $\{\lambda_{nN}\}$ decreases in N , we get

$$\sum_n |I_n| \log \frac{1}{\lambda_n} \leq A_4 .$$

That is the required property 2. The property 1 and the right hand inequality in 3 follow from (1.2.6). We are to check only the left hand inequality in 3. Let $v(t) = \lim_{N \to \infty} v_N(t)$ (the limit exists because $v_N(t)$) increases for fixed t and for growing N). Then (1.2.10) implies $v \in L^1(\partial \mathbb{D})$. Now using the property (1.1.1), we can write

$$\lambda_{n0} = \mu_n \geq c_2 \min(|I_n|, \frac{1}{|S'(\zeta_n)|}) \geq c_3 |I_n|^2 ;$$

and further due to (1.2.10)

$$a_{nN} \leq \frac{A_5}{|I_n|^2}$$

and then by induction we obtain

$$\lambda_{n,N+1} \geq \min\left(c_0 |I_n|^2, \frac{1}{A_5} |I_n|^2\right) \geq c_0 |I_n|^2,$$

where $c_0 = \min(c_3, 1/A_5)$ and that gives the left inequality in 3. The main lemma and Theorem 9 are proved. ●

§ 2. Multiplicity of a boundary zero of some classes of functions

If z_0, $z_0 \in \mathbb{D}$ is a zero of multiplicity N of an analytic function of in $\mathbb{D}$ then

$$\lim_{r \to 1-0} (1-r)^{-N} |f(rz_0)| \neq 0, \infty. \tag{2.1}$$

If $f \in C_A$, $z_0 \in \partial\mathbb{D}$, $f(z_0) = 0$ then there are two possible generalizations of the notion of the multiplicity of a zero. As a first way we may choose a scale of positive functions $h(r)$, $r \in (0,1)$, $h(r) \to 0$ for $r \to 1 - 0$ and consider

$$\overline{\lim_{r \to 1-0}} |f(rz_0)| / h(r) \tag{x}$$

similarly to (2.1). Then the zero z_0 has more multiplicity the faster the function decreases to zero as $r \to 1 - 0$ provided the limit in (x) is finite. As a second generalization of multiplicity we can consider something analogous to the number N . It is clear that such terms are closely connected with the Nevanlinna factorization of f at the vicinity of z_0 because if we introduce the atomic measure $d\nu_{z_0}=(1-|z_0|^2)\,\delta_{z_0}$; then f has at $z_0\in\mathbb{D}$ a zero of multiplicity N if and only if for any disc $D(z_0,\delta)$ we have

$$\int\limits_{f^{-1}(0)\cap D(z_0,\delta)} d\nu_{z_0} = N(1-|z_0|^2)$$

the integral being taken here over the set $f^{-1}(0)$ taking account of multiplicity. Hence if we associate with the Blaschke product with the inner and outer parts of f some measures in $\overline{\mathbb{D}}$ naturally related with them (see the definition in 2.2 below) then we may assume that the "multiplicity" of a zero $z_0\in\partial\mathbb{D}$ increases with the "intersivity" of these measures nearly z_0.

In this section we study some important classes of functions with rare or with fast decreasing Taylor-coefficients. There are restrictions in these classes on possible multiplicities of the boundary zeros compared with the classes Λ^n_ω or H^p_n because we can find functions $f\in A^\infty$ such that

$$f(z)\exp\frac{z+1}{1-z}\in A^\infty$$

and which have at the point 1 the boundary zero of the maximal possible multiplicity among the functions of Nevanlinna class N .

The problem of a possible rate of decreasing of $|f(r z_0)|$ to zero for functions with rare Taylor corfficients is closely connected with the problem of a possible rate of decreasing of Dirichlet series with rare exponents and was studied earlier (see the pioneere paper of L.Schwartz [19] or the paper of Hirshman and Jenkins [20]). But in our case with the help of another method we obtain in one situation an ultimate answer (part a) of Theorem 10) and in another situation we obtain for our problem the better result then that of Hirshman-Jenkins.

The question about local estimate of the singular measure for bounded functions with Hadamard gaps was posed and answered by J.M. Anderson [21] . Here in 2.2 we consider the general possible configuration of the factorization taking into account the outer part as well as the Blaschke product and obtain results analogous to that of J.M. Anderson. The problem of a possible multiplicity of a boundary zero for functions with fast decreasing Taylor coefficients is new. We answer this question in S.2.3.

2.0. Two lemmas.

DEFINITION. Let $f \in A$, $f(z) = \sum_{n\geq 0} \hat{f}(n) z^n$.
We say that f is (p, A) -lacunary, $p>1$, $A>0$ if there exists a sequence of exponents $\Lambda = \{n_m\}_{m\geq 0}$ such that $n_m \geq A m^p$ and $\hat{f}(n) = 0$ for $n \notin \Lambda$.

LEMMA 2.1. a) Let $A>0$, $p>2$,

$$\lambda_n = [An^p]+1, \quad \Lambda = \{\lambda_n\}_{n\geq 0},$$

$$\Pi_{p,A}(z) = \Pi_\Lambda(z) = \prod_{n=0}^{\infty} \left(1 - \frac{z}{\lambda_n}\right). \tag{2.0.1}$$

Let $\Gamma = \{\zeta = x+iy \in \mathbb{C} : x \geq -\frac{1}{2}, \quad |y| \leq 1\}$. Then if $z = re^{i\varphi} \in \mathbb{C}\setminus\Gamma$, $0<\varphi<2\pi$ then

$$\Pi_\Lambda(z) = \exp\left[\frac{\pi}{A^{1/p}\sin\frac{\pi}{p}} \cdot e^{i\frac{\varphi-\pi}{p}} r^{\frac{1}{p}} + O(\log r)\right] \tag{2.9.2}$$

$$|\Pi'_\Lambda(\lambda_n)| = \exp\left[c_p \frac{\lambda_n^{1/p}}{A^{1/p}} + O(\log \lambda_n)\right], \qquad \text{(2.0.3=}$$

where $c_p = \pi \operatorname{ctg} \frac{\pi}{p} > 0$.

b) Let $\lambda_n \geq An^p$, $\Lambda = \{\lambda_n\}_{n\geq 0}$ the function Π_Λ is defined in (2.0.1). Then for $p>2$ we have

$$|\Pi_\Lambda(-\xi)| \leq \exp\left[\frac{\pi}{A^{1/p}\sin\frac{\pi}{p}}\xi^{1/p} + C_0(A,p)\log(1+\xi)\right], \quad \xi>0, \tag{2.0.4}$$

and if $1<p\leq 2$, $\xi = re^{i(\pi+\theta)}$, $-\frac{\pi(p-q)}{2}\leq\theta\leq\frac{\pi(p-q)}{2}$, $q>0$, then

$$|\Pi_\Lambda(\xi)| \leq C_1(A,p,q)\exp\left[\left(\frac{\pi\cos\frac{\theta}{p}}{A^{1/p}\sin\frac{\pi}{p}}+\delta\right)r^{1/p}\right], \tag{2.0.5}$$

where $\delta=\delta(A,p,q)>0$ and $\delta\to 0$ as $q\to 0$.

PROOF is literally the same as that of relation (5.3_1) on p.94 of [35] ●

LEMMA 2.2. Let f be (p,A) -lacunary, $p>1$,
$f(z)=\sum_{n\geq 0}\hat{f}(n)z^n$, $\hat{f}(n)=0$, if $n\notin\Lambda=\{\lambda_n\}$, $f(x)\in L^1(0,1)$
and let

$$H_\rho(\zeta) \overset{def}{=} \sum_{n\geq 0}\frac{\hat{f}(n)\rho^n}{\zeta-n}, \quad 0<\rho<1, \quad \zeta\notin\mathbb{Z}_+.$$

Then for $\zeta\in\mathbb{C}\setminus\mathbb{Z}_+$ there exists the limit

$$H(\zeta) \overset{def}{=} \lim_{\rho\to 1-0} H_\rho(\zeta)$$

and H turns to be a meromorphic function bounded for $\operatorname{Re}\zeta\leq -1$ which can be represented in the form $H=\varphi/\Pi_\Lambda$ where φ is an entire function of order $1/p$ and

$$\Pi_\Lambda(\zeta)=\prod_n\left(1-\frac{\zeta}{\lambda_n}\right).$$

PROOF. Let $m=\int_0^1 \frac{|f(x)|}{\sqrt{x}}\,dx, \quad \rho=e^{-\delta},$ then for $\operatorname{Re}\zeta\leqslant -1$ we have

$$H_{e^{-\delta}}(\zeta)=\sum_{n=0}^{\infty}\frac{\hat{f}(n)\,e^{-\delta n}}{\zeta-n}=\int_{\delta}^{\infty} f(e^{-s-\delta})e^{s\zeta}\,ds$$

and for $\operatorname{Re}\zeta\leqslant -\frac{1}{2}$ and for $\rho\geqslant\frac{1}{2}$ we can write

$$|H_\rho(\zeta)|\leqslant\int_0^{\infty}|f(e^{-s-\delta})|e^{-\frac{s}{2}}\,ds\leqslant 2\int_0^1\frac{|f(h)|}{\sqrt{h}}\,dh=2m. \tag{2.0.6}$$

The function $\psi_\delta=H_\rho\Pi_\Lambda$ is an entire function of order $\frac{1}{p}$ because H_ρ is a meromorphic functions of order $\frac{1}{p}$, Π_Λ is an entire function of order $\frac{1}{p}$ and because for $\operatorname{Re}\zeta\leqslant-\frac{1}{2}$ and $\rho\geqslant\frac{1}{2}$ we have

$$|\psi_\delta(\zeta)|\leqslant 2m\,|\Pi_\Lambda(\zeta)|.$$

Let Ψ be an outer function in the half-plain $\{\operatorname{Re}\zeta\geqslant-1\}$ such that $|\Psi(\zeta)|=|\Pi_\Lambda(\zeta)|$ for $\operatorname{Re}\zeta=-1$. Then for a number $A=A_\Lambda>0$ we have

$$|\Psi(\zeta)|=|\Pi_\Lambda(\zeta)|\leqslant\exp A|\zeta|^{\frac{1}{p}},\quad \operatorname{Re}\zeta=-1,$$

and that implies

$$|\Psi(\zeta)|\leqslant\exp(A_1(|\zeta|+1)^{1/p}),\quad \operatorname{Re}\zeta\geqslant-1,$$

i.e. the order of Ψ in the half-plain $\{\operatorname{Re}\zeta\geqslant-1\}$ does not exceed $\frac{1}{p}$. Let $v_\delta=\psi_\delta/\Psi$. We have for $\operatorname{Re}\zeta=-1$:

$$|v_\delta(\zeta)| = |H_\rho(\zeta)| \cdot \left|\frac{\Pi_\Lambda(\zeta)}{\Psi(\zeta)}\right| \leqslant 2m$$

hence the Fragmen-Lindelöf principle applied to the half-plain $\mathrm{Re}\,\zeta \geqslant -1$ and to the function v_δ of order $\frac{1}{\rho}$ in that half-plain yields

$$|v_\delta(\zeta)| \leqslant 2m, \quad \mathrm{Re}\,\zeta \geqslant -1. \tag{2.0.7}$$

Let us choose now converging subsequences from the normal families of functions $H_\rho(\zeta)$ for $\mathrm{Re}\,\zeta < -\frac{1}{2}$ and of functions $v_\delta(\zeta)$ for $\mathrm{Re}\,\zeta \geqslant -1$ which is possible in view of (2.0.6) and (2.0.7). Let

$$H(\zeta) = \int_0^\infty f(e^{-s}) e^{s\zeta} d\zeta, \quad \mathrm{Re}\,\zeta \leqslant -\frac{1}{2}.$$

It is clear that $H_\rho(\zeta) \xrightarrow[\rho \to 1-0]{} H(\zeta)$, $\mathrm{Re}\,\zeta \leqslant -\frac{1}{2}$. Let v be a limit function for the subsequence v_{δ_n}. Then $|v(\zeta)| \leqslant 2m$ for $\mathrm{Re}\,\zeta \geqslant -1$. Furthermore, for $-1 < \mathrm{Re}\,\zeta < -\frac{1}{2}$ we have

$$\Pi_\Lambda(\zeta) H_\rho(\zeta) = v_\delta(\zeta) \Psi(\zeta),$$

hence $\Pi_\Lambda(\zeta) H(\zeta) = v(\zeta) \Psi(\zeta)$, $-1 < \mathrm{Re}\,\zeta < -\frac{1}{2}$.

We define a function φ as follows

$$\varphi(\zeta) = \begin{cases} \Pi_\Lambda(\zeta) H(\zeta), & \mathrm{Re}\,\zeta < -\frac{1}{2}, \\ v(\zeta)\Psi(\zeta), & \mathrm{Re}\,\zeta > -1. \end{cases}$$

The function φ is an entire function of order $\frac{1}{\rho}$ and $H = \varphi / \Pi_\Lambda$. ●

2.1. The radial decrease of (P, A)-lacunary functions.

In this section we deal with a possible rate of radial decrease of (P,A) -lacunary functions. For $p>2$ we find the answer up to the factor $O(1-r)^N$ and for $p<2$ we prove the result better than that, which can be obtained by application of Hirshman and Jenkins theorem [21] to (P,A) -lacunary functions. We point out that the constant $B_1(P,A)$ which is the best possible for $p>2$ fails to be such for $p<2$. However, we do not prove that here.

THEOREM 10. Let f be a (P,A) -lacunary function, $p>1$, $A>0$, $f \not\equiv 0$. Then

a) if $p>2$ then for a number $C_1 = C_1(P,A) > 0$ we have

$$\overline{\lim_{x\to 1-0}} |f(x)| \exp[B_1(1-x)^{-\frac{1}{p-1}} + C_1|\log(1-x)|] = +\infty, \tag{2.1.1}$$

where

$$B_1 = B_1(p,A) = (p-1)\left(\frac{\pi}{p\sin\frac{\pi}{p}}\right)^{\frac{p}{p-1}} A^{-\frac{1}{p-1}}$$

there exists a (P,A) -lacunary function $f_{P,A}$ such that

$$\lim_{x\to 1-0} |f_{P,A}(x)| \exp[B_1(1-x)^{-\frac{1}{p-1}} - C_1|\log(1-x)|] = 0. \tag{2.1.2}$$

b) if $1<p\leqslant 2$ then for any $\varepsilon>0$ we have

$$\overline{\lim_{x\to 1-0}} |f(x)| \exp[(B_2+\varepsilon)(1-x)^{-\frac{1}{p-1}}] = \infty, \tag{2.1.3}$$

where

$$B_2 = B_2(p,A) = B_1(p,A)\cdot\frac{(\cos\pi\frac{2-p}{p})^{\frac{p}{p-1}}}{(\cos\pi\frac{2-p}{p})^{\frac{1}{p-1}}}.$$

PROOF. Let

$$\tilde{f}(\xi)=\int_0^\infty f(e^{-s})e^{s\xi}\,ds.$$

We consider first the case $p>2$. Assume that for a number $C_1>0$ the limit in (2.1.1) is finite. Then

$$|\tilde{f}(\xi)|\leqslant c_2\int_0^\infty \exp\left[-B_1(1-e^{-s})^{-\frac{1}{p-1}}-C_1\log\frac{1}{1-e^{-s}}-s|\xi|\right]ds\leqslant$$

$$\leqslant c_2\int_0^\infty \exp\left[-B_1 s^{-\frac{1}{p-1}}-C_3\log\left(s+\frac{1}{s}\right)-s|\xi|\right]ds \stackrel{def}{=} c_2(I(\xi),\ \ \xi<0. \tag{2.1.4}$$

The derivative of the function $h(s)=B_1 s^{-\frac{1}{p-1}}+s|\xi|$ vanishes for $s_0=\left(\frac{B_1}{p-1}\right)^{\frac{p-1}{p}}\cdot|\xi|^{-\frac{p-1}{p}}$ hence applying to the integral $I(\xi)$ the classical asymptotic methods we obtain

$$|I(\xi)|\leqslant c_2'\exp\left[-B_1^{\frac{p-1}{p}}\cdot\frac{p}{(p-1)^{\frac{p-1}{p}}}\cdot|\xi|^{1/p}-C_4\log|\xi|\right],\ \ \xi<0.$$

where $C_4=C_4(p,B_1,C_3)>0$ for C_3 sufficiently large, $C_4\longrightarrow\infty$ as $C_3\longrightarrow\infty$. Hence if

$$\Pi_f(\zeta)=\prod_{n\geqslant 0:\hat{f}(n)\neq 0}\left(1-\frac{\zeta}{n}\right),$$

then Lemma 2.1 would imply for $\xi<0$ and for a large C_3 :

$$|\tilde{f}(\xi)\Pi_f(\xi)|\leqslant e^{-C_5(\log|\xi|+1)},\ \ C_5>0. \tag{2.1.5}$$

Because the function $\varphi = \tilde{f}\,\Pi_f$ is an entire function of order $\frac{1}{p}$ due to Lemma 2.2 Wiman's theorem and (2.1.5) yield $f \equiv 0$ for $p > 2$.

Assume now that $1 < p \leq 2$. We put $\theta_{1,2} = \pm \frac{\pi p}{2}$, $\xi = re^{i\theta_j}$, $j = 1,2$. As in (2.1.4) if the limit in (2.1.3) is finite we get

$$|\tilde{f}(\xi)| \leq c_6 \int_0^\infty \exp[-(B_2+\varepsilon)s^{-\frac{1}{p-1}} + rs\cos\frac{\pi p}{2}]\, ds \overset{def}{=} c_6 I_1(r)$$

and using the Laplace method for estimating of I_1 we obtain

$$I_1(r) \leq c_7 \exp\left[-\left(B_2+\frac{\varepsilon}{2}\right)^{\frac{p-1}{p}} \left|\cos\frac{\pi p}{2}\right|^{1/p} \cdot \frac{p}{(p-1)^{\frac{p-1}{p}}} r^{1/p}\right]. \tag{2.1.6}$$

Lemma 2.1, 2.2 and (2.1.6) yield that the entire function $\varphi = \tilde{f}\,\Pi_f$ of order $\frac{1}{p}$ is bounded on the rays $R_j = \{\zeta = re^{i\theta_j}\}$, $j = 1,2$ and then the uniqueness theorem in [40] implies

$$\int_{R_1 \cup R_2} \frac{\log|\varphi(\xi)|}{1+|\xi|^{1+1/p}}\, |d\xi| < \infty. \tag{2.1.7}$$

But then for a given $\varepsilon/2$ and suitable small δ (2.1.6) and (2.0.5) would imply the divergence of the integral (2.1.7). Hence the limit in (2.1.3) is infinite.

We have only to check (2.1.2). The required function $f_{p,A}$ is the following. Let

$$\lambda_n = [An^p] + 1,$$

$$\Pi_{p,A}(\zeta) = \prod_{n=0}^{\infty} \left(1 - \frac{\zeta}{\lambda_n}\right).$$

We put

$$f(z)=\sum_{n\geq 0}\frac{z^{\lambda^n}}{\Pi'_{P,A}(\lambda_n)},\quad z\in\mathbb{D}.$$

Let us check (2.1.2). First due to (2.0.1) - (2.0.3) we can write

$$f(e^{-s})=\frac{1}{2\pi i}\int_{-\infty i-\ell}^{\infty i-\ell} e^{-s\zeta}\,\frac{d\zeta}{\Pi_{P,A}(\zeta)},\qquad s>0,\ \ell\geq 1,$$

hence

$$|f(e^{-s})|\leq c\int_{-\infty i-\ell}^{\infty i-\ell} e^{s\ell}\cdot e^{-\alpha r^{1/P}\cos\frac{\theta}{P}+c_1\log r}\,dy, \tag{2.1.8}$$

where $\zeta=-\ell+iy=re^{i(\pi-\theta)}$, $\quad -\pi<\theta<\pi,\qquad \alpha=\dfrac{\pi}{A^{1/P}\sin\frac{\pi}{P}}$.
Further $\theta=\operatorname{arctg}\frac{y}{\ell}$ and if $\rho=1/P$

$$h(y)=r^{\rho}\cos\rho\theta=(\ell^2+y^2)^{\rho/2}\cos(\rho\operatorname{arctg}\tfrac{y}{\ell}),$$

then

$$h'(y)=\rho r^{\rho-1}\sin(1-\rho)\theta\geq 0,\qquad \theta\geq 0$$

and because h is even, we have $h(y)\leq h(0)$. If $y\geq\ell$ then $\theta\geq\frac{\pi}{4}$ hence

$$h'(y)\geq\frac{2}{\pi}\rho r^{\rho-1}(1-\rho)\theta\geq C_\rho y^{\rho-1},$$

$$h(y)-h(0)\geq\int_{\ell}^{y}h'\geq c_{\rho_1}y^{\rho},\quad y\geq 2\ell,$$

and then

$$\int_{-\infty}^{\infty}e^{-\alpha r^{\rho}\cos\rho\theta+c\log r}\,dy=\int_{-2\ell}^{2\ell}+\int_{|y|\geq 2\ell}\quad\leq$$

$$\leqslant c_2 l \cdot e^{-\alpha l^\rho + c_1 \log l} + c_3 \cdot e^{-\alpha l^\rho} \cdot \int_0^\infty e^{-c_{p_2} y^\rho} \leqslant e^{-\alpha l^\rho + c_4 \log l},$$

and due to (2.1.8)

$$|f_{P,A}(e^{-s})| \leqslant \min_{l \geqslant 1} \exp(sl - \alpha l^\rho + c_4 \log l). \tag{2.1.9}$$

Putting $l = (\alpha\rho)^{\frac{1}{1-\rho}} \cdot s^{\frac{1}{\rho-1}}$ in (2.1.9) we get (2.1.2). Theorem 10 is now proved. ●

2.2. Possible intensity of the factorization terms nearby a boundary zero.

In this section we introduce a characteristic of intensity of the factorization terms near a boundary point which is an analogue of multiplicity of an inner zero. The characteristic introduced can not take big values for all (P, A) .lacunary functions (Theorem 11); for functions with Hadamard gaps a weaker result, taking into account only the inner factor, was earlier obtained by J.M.Anderson [25]. Moreover if this characteristic is "almost extremal" then it must behave sufficiently regularly (Theorem 12).

DEFINITION. Let $f \in N$, $f = {}_e f B_f S_f$, μ be a singular measure corresponding to S_f, $\mu = \mu^+ - \mu^-$ measures μ^+ and μ^- being nonnegative, $Z = B_f^{-1}(0)$. We put

$$N_f(\zeta, \rho) \stackrel{def}{=} \int_{-2\arcsin\frac{\rho}{2}}^{2\arcsin\frac{\rho}{2}} \log^+ \frac{1}{|f(\zeta e^{i\alpha})|} d\alpha + \int_{-2\arcsin\frac{\rho}{2}}^{2\arcsin\frac{\rho}{2}} d\mu^+(\zeta e^{i\alpha}) + \frac{1}{2} \sum_{\substack{\alpha \in Z_f \\ |\alpha - \zeta| < \rho}} (1 - |\alpha|)^2, \ \zeta \in \partial\mathbb{D}.$$

THEOREM 11. a) Let f be (P, A) .lacunary, $p \geqslant 2$, $f \in H^q$ for $q > 0$. Then

$$\varlimsup_{\rho \to 0} \rho^{\frac{2-p}{p-1}} N_f(\zeta, \rho) < \infty, \quad \zeta \in \partial\mathbb{D}; \tag{2.2.1}$$

b) There exists a $(2,A)$-lacunary function f_A, $|f_A|>1$ such that $N_f(1,\rho)\geq c>0$ for all $\rho>0$.

c) For $p>2$ there exists a (P,A)-lacunary function $f_{P,A}$, $|f_{P,A}|<1$ such that

$$\varliminf_{\rho\to 0} \rho^{\frac{2-p}{p-1}} \int_{-\rho}^{\rho} \log \frac{1}{|f_{P,A}(e^{i\theta})|}\, d\theta > 0 \tag{2.2.2}$$

and

$$\varliminf_{\rho\to 0} \rho^{\frac{2-p}{p-1}} \sum_{|\alpha-1|+\rho|<\frac{\rho}{2}} (1-|\alpha|^2) > 0. \tag{2.2.3}$$

THEOREM 12. Let f be (P,A)-lacunary, $p>2$, $f\in H^q$ for $q>0$ and assume that

$$\varliminf_{\rho\to 0} \rho^{\frac{2-p}{p-1}} N_f(\zeta,\rho) > 0. \tag{2.2.4}$$

Then

$$\varlimsup_{\rho\to 0} \rho^{\frac{2-p}{p-1}} N_f(\zeta,\rho) < \infty. \tag{2.2.5}$$

PROOF OF THEOREM 11.

PART a) We assume $\zeta = 1$ and that $\int_{\partial\mathbb{D}} |f|^q = 1$ then

$$\frac{1}{2q\delta}\int_{-\delta}^{\delta} \log^+|f|^q \leq \frac{1}{q}\log\frac{1}{2\delta}\int_{-\delta}^{\delta}(1+|f|^q) \leq \frac{1}{q}\log\frac{2\pi+1}{2\delta},$$

i.e.

$$\int_{-\delta}^{\delta} \log^+|f| \leq c_0\,\delta\log\frac{1}{\delta},$$

then integrating by parts we have

$$\int\limits_{\partial\mathbb{D}} \log^+|f(\zeta)| \frac{1-x^2}{|\zeta-x|^2}|d(\zeta)| \leq \log\frac{1}{\rho}, \quad 0<x<1, \quad \rho=1-x. \tag{2.2.6}$$

The relation (2.2.6) is the first step in obtaining on upper bound for $|f(x)|$, $0<x<1$ which we are going to prove assuming that (2.2.1) is false. This estimate required will contradict to Theorem 10 and thus our assumption that (2.2.1) is false is wrong. Hence part a) of Theorem 11 will follow.

So assume that the limit in (2.2.1) is smaller that a number $C(P,A)$ the exact value of which we choose later. Since

$$\left|\frac{z-\zeta}{1-z\bar{\zeta}}\right|^2 = 1-\frac{(1-|z|^2)(1-|\zeta|^2)}{|1-z\bar{\zeta}|} \leq \exp\left[-\frac{(1-|z|^2)(1-|\zeta|^2)}{|1-z\bar{\zeta}|^2}\right], \quad z,\zeta\in\mathbb{D},$$

we have

$$|f(x)| \leq \exp\left(-\int\limits_{\mathbb{D}} \frac{1-x^2}{|1-z\bar{\zeta}|^2}\, d\mu(\zeta)\right)\cdot\exp\left(\int\limits_{\partial\mathbb{D}} \log^+|f(\zeta)| \frac{1-x^2}{|\zeta-x^2|}\, |d\zeta|\right), \tag{2.2.7}$$

where the measure μ is defined in the closed disc $\overline{\mathbb{D}}$ as follows:

$$d\mu(\zeta) = \begin{cases} \frac{1}{2}(1-|\zeta|^2)\delta_\zeta, & \zeta\in Z_f(\mathbb{D}), \\ 0, & \zeta\in\mathbb{D}\setminus Z_f(\mathbb{D}), \\ d\mu_s^+(\zeta)+\log^+\frac{1}{|f(\zeta)|}, & \zeta\in\partial\mathbb{D}. \end{cases}$$

$d\mu_s^+$ denotes the nonnegative singular measure corresponding to the measure S_f. We note that

$$\mu\left(\overline{D(1,\rho)}\right) = N_f(1,\rho). \tag{2.2.8}$$

and then (2.2.6) and (2.2.7) yield

$$|f(x)|\leq \tilde{c}\rho^{-c_0}\exp\left(-\frac{1}{2\rho}N_f(1,\rho)\right),\quad 0<x<1,\ \rho=1-x. \tag{2.2.9}$$

Theorem 10 implies

$$\overline{\lim_{x\to 1-0}}\,|f(x)|\exp(2B(1-x)^{-\frac{1}{p-1}})=+\infty,$$

where $B=B_1(P,A)$. If we would have $N_f(1,\rho)\geq 5B\rho^{\frac{p-2}{p-1}}$ for all $\rho\leq\rho_0$ then (2.2.9) would yield

$$\lim_{x\to 1-0}|f(x)|\exp(2B(1-x)^{-\frac{1}{p-1}})=0$$

hence we would obtain finally

$$\underline{\lim}_{\rho\to 0}\,\rho^{\frac{2-p}{p-1}}N_f(1,\rho)\leq 5B_1(p,A).$$

PART b) The function taken from [50] is $(2,A)$ -lacunary and its singular measure contains an atomic measure at the point 1; then (2.2.8) implies

$$N_f(1,\rho)=\mu(\overline{D(1,\rho)})\geq\mu_s^+(\{1\})=c_0>0.$$

PART c) We omit some technical standard details when dealing with the asymptotic. Let

$$A_1=a_pA,\quad a_p=\left(\frac{2\cos\frac{\pi}{p}}{\cos\frac{3\pi}{4p}}\right),\quad \lambda_n=[An^p]+1,$$

$$\Pi_0(\zeta)=\Pi_{p,A_1}(e^{\frac{3\pi i}{4}}\zeta),\quad \Pi_1(\zeta)=\Pi_{p,A_1}(e^{-\frac{3\pi i}{4}}\zeta),$$

$$\Pi=\Pi_{p,A}\,.$$

We put

$$\hat{f}_j(m)=\begin{cases}0, & m\neq\lambda_n\\ \dfrac{\Pi_j(\lambda_n)}{\Pi'(\lambda_n)}, & m=\lambda_n,\end{cases}$$

$$f_j(z)=\sum_{m\geq 0}\hat{f}_j(m)z^m, \qquad z\in\mathbb{D}, \quad j=0,1,$$
$$f=f_0+f_1. \tag{2.2.10}$$

The function f is the required one. To check (2.2.2) it is sufficient to check that

$$|f_j(e^{ix})|\leq C\exp(-b|x|^{-\frac{1}{p-1}}), \quad j=0,1, \quad |x|\leq\varepsilon_0.$$

Lemma 2.1 yields

$$|\hat{f}_j(\lambda_n)|\leq c\exp\left(\frac{C_p}{2A^{1/p}}\lambda_n^{1/p}\right)\cdot\exp\left(-\frac{C_p}{A^{1/p}}\lambda_n^{1/p}\right)\leq Ce^{-c_1 n}. \tag{2.2.11}$$

Let us consider the contour $\Gamma=\{\zeta=x+iy\in\mathbb{C}: x=-1, \quad |y|\leq 1\}\cup$ $\cup\{y=1, x\geq -1\}\cup\{y=-1, \quad x\geq -1\}$ and let $\varphi_j=\Pi_j/\Pi$, $j=0,1$. By Lemma 2.1 we have

$$|\varphi_j(x+iy)|\leq C\exp\left[-\frac{\pi\,\mathrm{ctg}\,\frac{\pi}{p}}{A^{1/p}}x^{1/p}+\frac{\pi\cos\frac{3\pi}{4p}}{A_1^{1/p}\sin\frac{\pi}{p}}x^{1/p}+O(\log x)\right]\leq$$

$$\leq ce^{-c_3x^{1/p}}, \quad y_0=\pm 1, \quad x\geq 1, \quad j=0,1,$$

hence

$$\int_\Gamma|\varphi_j(\zeta)||d\zeta|<\infty, \quad j=0,1 \tag{2.2.12}$$

and using (2.2.11) and (2.2.12) we write

$$f_j(e^{ix}) = \sum_{n\geq 0} \hat{f}_j(n) e^{inx} = \frac{1}{2\pi i}\int_\Gamma e^{ixz}\,\varphi_j(z)dz. \tag{2.2.13}$$

Using Lemma 2.1 again we deduce from (2.2.13) that

$$f_j(e^{ix}) = -\frac{1}{2\pi i}\int_{-\infty+i}^{\infty+i} e^{ixz}\,\varphi_j(z)\,dz + \frac{1}{2\pi i}\int_{-\infty-i}^{\infty-i} e^{ixz}\,\varphi_j(z)\,dz.$$

We consider only the case $x > 0$; the case $x < 0$ is analogous .

For $\mathrm{Im}\, z \geq 0$ we have $|e^{ixz}| \leq 1$ hence

$$f_j(e^{ix}) = \frac{1}{2\pi i}\int_{-\infty-i}^{\infty-i} e^{ixz}\,\varphi_j(z)dz = \int_{-\infty-il}^{\infty-il} e^{ixz}\,\varphi_j(z), dz, \tag{2.2.14}$$

where $l \geq 1$ is arbitrary. Using Lemma 2.1 as in (2.1.8) - (2.1.9) we get

$$|f_j(e^{ix})| \leq e^{xl}\int_{-\infty-il}^{\infty-il} e^{-c|z|^{1/p}}\,|dz| \leq c_1 e^{xl - c_2 l^{1/p}}$$

and for $l = \left(\frac{c_2}{p}\right)^{\frac{p}{p-1}} x^{-\frac{p}{p-1}}$ we obtain $|f_j(e^{ix})| \leq c_4 e^{-c_5 x^{-\frac{1}{p-1}}}$,

i.e.

$$|f(e^{ix})| \leq |f_0(e^{ix})| + |f_1(e^{ix})| \leq 2c_4 e^{-c_5 x^{-\frac{1}{p-1}}},$$

and that is the required (2.2.2).

We check now that the function f has sufficiently many zeros on $(0,1)$. We write for brevity $a \overset{\log}{\approx} b$ if $\left|\log\frac{a(s)}{b(s)}\right| \leq c\log s$, $s > s_0 > e$ for a constant C . Using formula (2.3.2) for the asymptotic of Π_j and arguing in a standard way, we obtain

$$f_j(e^{-s}) = -\frac{1}{2\pi i}\int_{-\infty i-l}^{\infty i-l} e^{-sz}\psi_j(z)\,dz \overset{\log}{\sim}$$

$$\overset{\log}{\sim} c\int_{-\infty i-l}^{\infty i-l} \exp[-sz-(a\pm bi)z^{1/p}]\,dz \overset{\log}{\sim} \exp[-(a_1\pm b_1 i)s^{-\frac{1}{p-1}}] \tag{2.2.15}$$

with some real numbers $a_1>0,\ b_1\neq 0$ and it must be taken into account that in the formula (2.2.15) the values b_1 and $-b_1$ correspond to f_0 and f_1. The definition of f_j implies $f_0(e^{-s}) = \overline{f_1(e^{-s})},\ 0<s<\infty$. Hence the asymptotic (2.2.15) yields

$$f(e^{-s}) = \exp[-a_1 s^{-\frac{1}{p-1}} + h_1(s)]\cdot\cos(b_1 s^{-\frac{1}{p-1}} + h_2(s)), \tag{2.2.16}$$

where $|h_j(s)|\leq c\log(1+1/s)$. Let $\{s_n\}$ be a sequence of decreasing numbers defined from the equalities

$$|b_1|s_n^{-\frac{1}{p-1}} = \pi n.$$

Then (2.2.16) implies that for $n\geq n_0$ we have $f(e^{-s_{2n}})>0$, $f(e^{-s_{2n+1}})<0$ hence any interval (s_{2n+1}, s_{2n}) contains a zero of f. The definition of s_n implies (2.2.3). Theorem 11 is proved.

PROOF OF THEOREM 12. Let

$$\varepsilon_0 = \underline{\lim}_{\delta\to 0}\ N_f(1,\delta)\,\delta^{\frac{2-p}{p-1}} > 0. \tag{2.2.17}$$

Assume that

$$\overline{\lim}_{\delta\to 0}\ N_f(1,\delta)\,\delta^{\frac{2-p}{p-1}} \geq B, \tag{2.2.18}$$

where we define the number B below and then deduce a contradiction which will finish the proof so suppose that for a $\delta > 0$ we have

$$\delta > 0 \quad N_f(1,\delta) > B_1 \delta^{\frac{p-2}{p-1}}. \tag{x}$$

We put $s_0 = \frac{2B_1}{p-1}\delta^{-\frac{p}{p-1}}$ and estimate the function $\tilde{f}(-s)$ for $s_0 \leqslant s \leqslant 2s_0$, $s_0 > 0$. Using (2.2.17) and our assumption (x) we obtain as in (2.1.14) the following estimates

$$|f(e^{-x})| \leqslant \begin{cases} C_* \cdot \exp(-c_1 \varepsilon_0 x^{-\frac{1}{p-1}}), & x > 0, \\ C_* \cdot \exp(-c_3 B_1 \delta^{-\frac{1}{p-1}}), & c_2 \delta \leqslant x \leqslant \delta, \end{cases} \tag{2.2.19}$$

where c_1 is an absolute constant, c_3 depends only on c_2, $c_3 = c_1 c_2$. Let $c_2 = \left(\frac{c_1 \varepsilon_0}{2B_1}\right)^{\frac{p-1}{p}}$. Then (2.2.19) yields

$$|\tilde{f}(-s)| \leqslant c_4(\varepsilon_0) \exp(-c_5(\varepsilon_0) B_1^{\frac{p-1}{p}} s_0^{1/p}), \qquad s_0 \leqslant s \leqslant 2s_0,$$

hence

$$\int_{s_0}^{2s_0} \log|\tilde{f}(-s)|\,ds \leqslant -B_2 s_0^{\frac{1}{p}+1} + C_6(\varepsilon_0) s_0, \tag{2.2.20}$$

where $B_2 = C_5(\varepsilon_0) B_1^{\frac{p-1}{p^2}}$. But by Lemma 2.2 $\tilde{f} = \psi/\Pi_f$, ψ is an entire function of order $\frac{1}{p}$ and of the type not exceeding $C_7 = C_7(p, A, c_*, q)$. Let the Taylor expansion of ψ have the form $\psi(\zeta) = a\zeta^N + \ldots$, $n_\psi(r)$ be the number of zeros of ψ in $D(0,r)$. Then [36] for $r \geqslant 1$ we have

$$n_\psi(r) \le c_8\left(\frac{1}{\rho}, c_7, a, N\right) r^{\frac{1}{\rho}} = c_8(\psi) r^{\frac{1}{\rho}}. \qquad (2.2.21)$$

Further, because $\left|1-\frac{z}{\mu}\right| \geq \left|1-\left|\frac{z}{\mu}\right|\right|$ and all zeros of Π_f lie on $[0,\infty)$ we have

$$\log|\tilde{f}(-s)| = \log|\psi(-s)| - \log|\Pi_f(-s)| \geq$$

$$\geq \int_0^\infty \log\left|1-\frac{s}{t}\right| dn_\psi(t) - \int_0^\infty \log\left(1+\frac{s}{t}\right) dn_{\Pi_f}(t) - C_{\Pi_f}\log s. \qquad (2.2.22)$$

Let $n(t) = n_\psi(t) + n_{\Pi_f}(t)$. Continuing (2.2.22), we obtain with account of (2.2.21) that

$$\log|\tilde{f}(-s)| \geq \int_0^{4s_0} \log\left|1-\frac{s}{t}\right| dn_\psi(t) - \int_0^{4s_0} \log\left(1+\frac{s}{t}\right) dn_{\Pi_f}(t) -$$

$$-c_0 s \int_{4s_0}^\infty \frac{dn(t)}{t} - C_{\Pi_f}\log s \geq \int_0^{4s_0} \log\left|1-\frac{s}{t}\right| dn_\psi(t) - \int_0^{4s_0} \log\left(1-\frac{s}{t}\right) dn_{\Pi_f}(t) -$$

$$-c_{10}\, s_0^{1/\rho}, \qquad c_{10} = c_{10}(\psi, \Pi_f), \qquad dn(0) = 0,$$

hence

$$\int_{s_0}^{2s_0} \log|\tilde{f}(-s)|\,ds \geq \int_0^{4s_0} dn_\psi(t) \int_{s_0}^{2s_0} \log\left|1-\frac{s}{t}\right| ds -$$

$$-\int_0^{4s_0} dn_{\Pi_f}(t) \int_{s_0}^{2s_0} \log\left(1+\frac{s}{t}\right) ds - c_{11} s^{\frac{1}{\rho}+1} \geq -c_{12}\, s_0^{\frac{1}{\rho}+1}, \qquad (2.2.23)$$

where $c_{12} = c_{12}(\psi, \Pi_f)$ because due to $dn(0) = 0$ we have

$$\int_0^{s_0/2} s_0 \log \frac{2s_0}{t}\, dn(t) \leq c_{13}(\psi, \Pi_f)\, s_0^{1+\frac{1}{\rho}}$$

and finally because

$$\int_{s_0}^{2s_0} |\log|1-\frac{s}{t}||\, ds \leq \text{const}\cdot s_0, \quad \frac{s_0}{2} \leq |t| \leq 4s_0.$$

Due to relations (2.2.20) and (2.2.23) we obtain $B_2 \leq 2c_{12}$ for sufficiently large s_0 and then $B_1 \leq c_{14}(\psi, \Pi_f)$ hence the assumption $B \geq 2c_{14}$ leads to the contradiction. Theorem 12 now proved.

2.3. Radial decreasing of functions with fast decreasing Taylor coefficients.

In this section we prove that if a function with Taylor coefficients is bounded from above by rapidly decreasing majorant and by the majorant $\exp(-b \log^{\lambda} \frac{1}{1-x})$, $\lambda > 1$, then it has, in fact, a smaller majorant related to the rate of decrease of the Taylor coefficients. This result is ultimate in a sence; for example, there is no connection between the rates of decreasing of Taylor coefficients and that of the function along the radius if the majorant equals $\exp(-b \log \frac{1}{1-x})$ because the function

$$(1-x)^M = \exp(-M \log \frac{1}{1-x})$$

has the required rate of decreasing along $(0,1)$ and its Taylor coefficients are zero starting from some place.

THEOREM 13. Let f be analytical in $\mathbb{D}$ and let

$$|f(x)| < c \exp(-b \log^{\lambda} \frac{1}{1-x}), \quad \lambda > 0, \quad b > 0, \quad 0 < x < 1,$$

Then

a) if $|\hat{f}(n)| < c_f \exp(-\sigma\sqrt{n})$, then $f \equiv 0$ (2.3.1)

b) if $|\hat{f}(n)| < c_f \exp(-\sigma n^p)$, $0 < p < \frac{1}{2}$, then

$$|f(x)| < c \exp\left(-b(1-x)^{-\frac{p}{1-p}}\right) \tag{2.3.2}$$

for a number $b > 0$.

c) for any p, $0 < p < \frac{1}{2}$ there exists a function $f_p \not\equiv 0$ such that

$$|f_p(x)| < c\left(-b(1-x)^{-\frac{p}{p-1}}\right), \quad |\hat{f}_p(n)| < c\exp(-\sigma n^p) \tag{2.3.3}$$

for some $b > 0$, $\sigma > 0$, $c > 0$.

PROOF.

PART a) First we observe that

$$|\tilde{f}(-s+it)| = \left|\int_0^\infty (e^{-x})\, e^{-xs+itx}\, dx\right| \leqslant$$

$$\leqslant c\int_0^\infty \exp\left[-b_1 \log^\lambda\left(\frac{1}{x}+1\right) - sx\right] dx \underset{b_1,\lambda,c}{\ll} \exp(-b_2 \log^\lambda s), \quad s > 10. \tag{2.3.4}$$

Let

$$h_N(\zeta) = \sum_{\nu=0}^{N} \frac{\hat{f}(\nu)}{\zeta - \nu}$$

and define R_N from the equality

$$b_2 \log^\lambda \frac{R_N}{2} = \frac{\sigma}{2}\sqrt{N}, \tag{2.3.5}$$

i.e.

$$R_N = 2\exp \sigma_1 N^{\frac{1}{2\lambda}}.$$

We also define the arcs Γ_N and T_N :

$$\Gamma_N = \{\zeta : |\zeta| = R_N,\ \operatorname{Re}\zeta \leqslant -\frac{R_N}{2}\},$$

$$T_N = \partial D(0, R_N) \setminus \Gamma_N.$$

The condition (2.3.1) yields

$$|\tilde{f}(\zeta) - h_N(\zeta)| \leqslant Ce^{-\frac{\sigma}{2}\sqrt{N}} \tag{2.3.6}$$

for $|\zeta - \nu| \geqslant 1$, $\nu = 0, 1, 2, \ldots$. The estimate (2.3.4) and the definition of R_N imply

$$|\tilde{f}(\zeta)| \leqslant ce^{-\frac{\sigma}{2}\sqrt{N}}, \quad \zeta \in \Gamma_N \tag{2.3.7}$$

and then from (2.3.6) and (2.3.7) we obtain

$$|h_N(\zeta)| \leqslant c_1 e^{-\frac{\sigma}{2}\sqrt{N}}, \quad \zeta \in \Gamma_N. \tag{2.3.8}$$

It is clear that

$$|h_N(\zeta)| \leqslant c_2, \quad \zeta \in T_N. \tag{2.3.9}$$

Because h_N is analytical in $\mathbb{C} \setminus D(0, R_N)$, $h(\infty) = 0$ the estimates (2.3.8) and (2.3.9) and standard arguments for outer functions imply

$$|h_N(\zeta)| < Ce^{-\sigma_2\sqrt{N}}, \quad |\zeta| = 2R_N. \tag{2.3.10}$$

We now apply the Hadamard three-circles theorem to the function h_N and to the circles $\gamma_1 = \partial D(0,2N)$, $\gamma_2 = \partial D(0,4N)$ and $\gamma_3 = \partial D(0,2R_N)$. On γ we have (2.3.10), in γ_1 we have $|h_N(\zeta)| \leqslant C$ hence for $\zeta \in \gamma_2$ we get

$$|h_N(\zeta)| \leqslant C_1\left[\exp(-\sigma_2\sqrt{N})\right]^{\frac{\log\frac{4N}{2N}}{\log\frac{2R_N}{2N}}} = C_1 \exp\left(-\sigma_2 N^{\frac{1}{2}} \frac{\log_2}{\log R_N - \log N}\right) \leqslant$$
$$\leqslant c\exp\left(-\sigma_3 N^{\frac{1}{2}-\frac{1}{2\lambda}}\right) = C\exp\left(-\sigma_3 N^{\mu}\right), \quad \frac{1}{2} > \mu > 0. \tag{2.3.11}$$

Taking (2.3.8) and (2.3.11) into account we see that on $\Gamma = \mathbb{C}\setminus L$, $L = \{\zeta = x+iy;\, x \geqslant -1, |y| \leqslant 1\}$ and on $\partial\Gamma$ holds

$$|\tilde{f}(\zeta)| \leqslant Ce^{-\sigma_4|\zeta|^{\mu}}, \quad \zeta \in \overline{\Gamma}. \tag{2.3.12}$$

We now are almost in the same position as at the beginning of the proof but we now have for the function $\tilde{f}$ the better estimates (2.3.12) instead of (2.3.11). We iterate the arguments (2.3.5) - (2.3.12) with the new estimate and obtain $\tilde{f} \equiv 0$, $f \equiv 0$.

We choose ρ_N now from the equality

$$\sigma_4 \rho_N^{\mu} = \frac{\sigma}{2} N^{1/2}, \tag{2.3.13}$$

i.e. $\rho_N = \sigma_5 N^{\frac{1}{2\mu}}$. Then for $\zeta \in \partial D(0,\rho_N) \cap \Gamma$ we have

$$|\tilde{f}(\zeta)| \leq Ce^{-\frac{\sigma}{2}N^{\frac{1}{2}}}.$$

Hence with account of (2.3.8) we get

$$|h_N(\zeta)| \leq c_1 e^{-\frac{\sigma}{2}N^{\frac{1}{2}}}, \quad \zeta \in \partial D(0,\rho_N) \cap \Gamma. \tag{2.3.14}$$

But now (2.3.14) and (2.3.9) imply similar to (2.3.10) that

$$|h_N(\zeta)| \leq c_2 e^{-\sigma_2 N^{\frac{1}{2}}}, \quad \zeta \in \partial D(0, 2\rho_N). \tag{2.3.15}$$

We use the Hadamard three-circles theorem again to h_N and to the circles $\gamma_1 = \partial D(0, 2N)$, $\gamma_2 = \partial D(0, 4N)$ and $\gamma_3 = \partial D(0, 2\rho_N)$. Then due to (2.3.15) we deduce

$$|h_N(\zeta)| \leq c_3 [\exp(-\sigma_2 N^{\frac{1}{2}})]^{\frac{\log 2}{\log\rho_N - \log N}} \leq c_3 \exp\left(-\tilde{\sigma}_3 \frac{N^{\frac{1}{2}}}{\log N}\right), \quad |\zeta| = 4N, \tag{2.3.16}$$

and then for $\zeta \in \partial\Gamma$ we get from (2.3.16)

$$|h_N(\zeta)| \leq Ce^{-\tilde{\tilde{\sigma}}_4 \frac{|\zeta|^{1/2}}{\log|\zeta|}},$$

$$\int_{\partial\Gamma} \frac{|\log|\tilde{f}(\zeta)||}{|\zeta|^{3/2}} |d\zeta| = \infty$$

which implies $\tilde{f} \equiv 0$, $f \equiv 0$.

PART b) We argue here as in part a) but more accurately. First as in (2.3.5) - (2.3.12) and (2.3.13) - (2.3.16) we obtain that $\tilde{f}$ has the following estimate:

$$|\tilde{f}(\zeta)| \leq c \exp\left(-\sigma_1 \frac{|\zeta|^p}{\log|\zeta|}\right), \quad \zeta \in \bar{\Gamma}. \tag{2.3.17}$$

We have to verify that as a matter of fact the function $\tilde{f}$ has better bound than in (2.3.17) . We put

$$M(r) \stackrel{def}{=} \max_{\substack{|\zeta|=x \\ \zeta \in \bar{\Gamma}}} |\tilde{f}(\zeta)|,$$

$$V(r) = -\frac{r^p}{\log M(r)} .$$

Then (2.3.17) yields

$$V(r) \leq \frac{1}{\sigma_1} \log r + c_1, \quad r \geq 1. \tag{2.3.18}$$

We use now the following elementary fact.

LEMMA 2.3. Let $A \in C[0,\infty)$, $A(x) \leq ax + b$, $a, b > 0$, $x > 0$, $A^*(x)$ be the least concave majorant of A , i.e.

$$A^*(x) \stackrel{def}{=} \inf_{\substack{\{u,v\}: A(y) \leq uy+v, \\ y \geq 0.}} (ux+v).$$

Then if A is unbounded for any $A_0 > 1$ and for any $x_* > 0$ there exists $x_0 > x_*$ such that

$$A(x_0) \geq \frac{3}{4} A^*(x_0) \geq A_0,$$

$$A^*(y) \leq 2A^*(x_0) \cdot \frac{y}{x_0}, \quad y \geq x_0. \quad \bullet$$

We finish the proof of part b). If the function V is bounded then we have an estimate

$$|\tilde{f}(\zeta)| \leqslant C\exp(-b|\zeta|^p), \quad \zeta \in \Gamma, \tag{2.3.18'}$$

with a number $b>0$ and then we can deduce from it as in (2.1.8) - (2.1.9) the required estimate (2.3.2). So assume that V is unbounded. Then the function $A(x) = V(e^x)$ is unbounded too and (2.3.18) imply

$$A(x) \leqslant \frac{1}{\sigma_1} x + C_1, \quad x \geqslant 0. \tag{2.3.19}$$

Let A^* be the least concave majorante of A. We can apply Lemma 2.3 to A^* due to (2.3.19). We choose the number x_0 sufficiently large and such that

$$A^*(y) \leqslant 2A^*(x_0)\cdot\frac{y}{x_0}, \quad y \geqslant x_0,$$

$$A(x_0) \geqslant \frac{3}{4} A^*(x_0).$$

Let $r_0 = e^{x_0}$, $R = r_0 V^{\frac{1}{p}}(r_0)$, $V^*(r) \stackrel{def}{=} A^*(\log r)$.

We have

$$\log\frac{1}{M(R)} = \frac{R^p}{V(R)} = \frac{r_0^p V(r_0)}{V(r_0 V^{1/p}(r_0))} \geqslant$$

$$\geqslant \frac{3}{4}\cdot\frac{r_0^p V^*(r_0)}{V^*(r_0 V^{1/p}(r_0))} \geqslant \frac{3}{4}\cdot\frac{r_0^p V(r_0)}{V^*(r_0^2)} =$$

$$= \frac{3}{4} r_0^p \frac{A^*(\log r_0)}{A^*(2\log r_0)} > \frac{1}{8} r_0^p, \tag{2.3.20}$$

in the case x_0 is sufficiently large for the inequality $V^{1/\rho}(r_0) \leqslant r_0$ (due to (2.3.18) it is possible). But if $\tilde{\gamma}_1 = \partial D(0,R) \cap \Gamma$, $\tilde{\gamma}_2 = \partial D(0,r_0) \cap \Gamma$, $\tilde{\gamma}_3 = \partial D(0,\frac{r_0}{2}) \cap \Gamma$, then for $\xi \in \tilde{\gamma}_j$, $j = 1,2,3$, we have

$$|\tilde{f}(\zeta) - h_{[\frac{r_0}{2}]-1}(\zeta)| < c\exp(-\sigma_2 r_0^\rho) \tag{2.3.21}$$

and taking (2.3.20) into account we find

$$|h_{[\frac{r_0}{2}]-1}(\zeta)| < C\exp(-\sigma_3 r_0^\rho), \quad \zeta \in \partial D(0,R) \cap \Gamma, \tag{2.3.22}$$

and then

$$|h_{[\frac{r_0}{2}]-1}(\zeta)| < C_1\exp(-\sigma_4 r_0^\rho), \quad \zeta \in \partial D(0,2R). \tag{2.3.23}$$

We use the Hadamard theorem again to $h_{[\frac{r_0}{2}]-1}$ and to the circles $\gamma_1 = \partial D(0,2R)$, $\gamma_2 = \partial D(0,r_0)$, $\gamma_3 = \partial D(0,r_0/2)$.

Because $|h(\zeta)| \leqslant C$, $\zeta \in \gamma_3$ (2.2.23) yields

$$|h_{[\frac{r_0}{2}]-1}(\zeta)| \leqslant C_2\left[\exp(-\sigma_4 r_0^\rho)\right]^{\frac{\log 2}{\log \frac{4R}{r_0}}} = C_2\exp\left(-\sigma_5 \frac{r_0^\rho}{\log V(r_0)}\right), \quad |\zeta| = r_0. \tag{2.3.24}$$

Hence for $\zeta \in \tilde{\gamma}_2 = \partial D(0,r_0) \cap \Gamma$ we obtain from (2.3.24) and (2.3.21) that

$$|\tilde{f}(\zeta)| \leqslant C_3\exp\left(-\sigma_5 \frac{r_0^\rho}{\log V(r_0)}\right). \tag{2.3.25}$$

Then (2.3.25) yields

$$M(r_0) \leq c_3 \exp\left(-\sigma_5 \frac{r_0^\rho}{\log V(r_0)}\right). \tag{2.3.26}$$

But due to definition of V we have

$$M(r_0) = \exp\left(-\frac{r_0^\rho}{V(r_0)}\right),$$

i.e.

$$\exp\left(-\frac{r_0^\rho}{V(r_0)}\right) \leq c_3 \exp\left(-\sigma_5 \frac{r_0^\rho}{\log V(r_0)}\right),$$

$$\sigma_5 \frac{r_0^\rho}{\log V(r_0)} \leq \frac{r_0^\rho}{V(r_0)} + c_4 \leq \frac{2r_0^\rho}{V(r_0)},$$

if r_0 is sufficiently large and then

$$V(r_0) \leq \frac{2}{\sigma_5} \log V(r_0). \tag{2.3.27}$$

Let A_0 be the root of the equation

$$A_0 = \frac{2}{\sigma_5} \log A_0$$

then (2.3.27) yields $V(r_0) \leq A_0$ but that contradicts the assumption about unboundedness of V. Hence V is bounded and then (2.3.18') completes the proof of part b).

PART c) The required function is for example the function $f_{\rho,1}$ from Theorem 10 where $P = \frac{1}{\rho}$. This follows from Theorem 10 and Lemma 2.1. Theorem 13 is now proved. ●

CHAPTER IV

CLOSED IDEALS IN THE SPACE $X^r_{pq}(\omega, \ell)$.

In present chapter we extend the list of Banach algebras of functions analytic in $\mathbb{D}$ smooth up to the boundary in which the structure of closed ideals in known. We describe all closed ideals in the algebras $X^r_{pq}(\omega, \ell)$ the special case of which for $p = q = \infty$, $\ell \equiv 1$ are algebras Λ^r_ω where all closed ideals were studied earlier [26] , [27] . The general and nowadays unique known method of describing the closed ideals in Banach algebras of analytic functions smooth up to the boundary makes use of the duality which reduces the description of the closed ideals to a special approximation problem. We consider below only the later question since the reduction of the ideal problem to the approximation problem is well known [23] , [24] , [27] . The integral norm in the space $X^r_{pq}(\omega, \ell)$ forced us to introduce an equivalent norm which we discuss in § 1. The approximation problem is considered in § 2.

DEFINITION 1. Let ω be a modulus of continuity, $\ell \geqslant 0$ be a function, $\ell \in L^1(\partial\mathbb{D})$, $\int_I \ell > 0$ for any arc $I \subset \partial\mathbb{D}$, p, q, r being numbers $r \in \mathbb{Z}_+$, $1 \leqslant p, q < \infty$. Let

$$\lambda(\theta; h) \overset{def}{=} \int_\theta^{\theta+h} \ell(e^{i\alpha})\, d\alpha.$$

We define the space $X = X^r_{pq}(\omega, \ell)$ as follows $f \in X \Leftrightarrow$

$\Leftrightarrow f \in C_A$ and

$$|f|_X \overset{\text{def}}{=} \left\{ \int_0^1 \frac{dh}{h} \left[\int_0^{2\pi} \left| \frac{f^{(r)}(e^{i(\theta+h)}) - f^{(r)}(e^{i\theta})}{\lambda(\theta;h)} \right|^p d\theta \right]^{q/p} \right\}^{1/q} < \infty. \tag{1}$$

If p or q equals infinity we modify the definition. If $p=\infty$, $q<\infty$ then

$$|f|_X \overset{\text{def}}{=} \left\{ \int_0^1 \frac{dh}{h} \sup_\theta \left| \frac{f^{(r)}(e^{i(\theta+h)}) - f^{(r)}(e^{i\theta})}{\lambda(\theta;h)} \right|^q \right\}^{1/q} < \infty;$$

if $p<\infty$, $q=\infty$, then

$$\mathcal{Y}_f(h) \overset{\text{def}}{=} \left[\int_0^{2\pi} \left| \frac{f^{(r)}(e^{i(\theta+h)}) - f^{(r)}(e^{i\theta})}{\lambda(\theta;h)} \right|^p d\theta \right]^{1/p} = o(1), \quad h \to 0;$$

if $p=\infty$, $q=\infty$, then

$$\mathcal{Y}_f(h) = \sup_\theta \left| \frac{f^{(r)}(e^{i(\theta+1)}) - f^{(r)}(e^{i\theta})}{\lambda(\theta;h)} \right| = o(1), \quad h \to 0, \quad |f|_X \overset{\text{def}}{=} \sup_{(0,1)} \mathcal{Y}_f(h).$$

We introduce the norm in the space $X = X^r_{pq}(\omega, l)$ as follows

$$\|f\|_X \overset{\text{def}}{=} |f|_X + \|P_r(f;0;\cdot)\|_{C_A}. \tag{2}$$

In what follows we consider only the case $p<\infty$, $q<\infty$ because the limit cases are simpler. The definition of the standard ideal is given in the Introduction.

THEOREM 14. Let a modulus of continuity ω satisfy the conditions

$$c_1\left(\frac{y}{x}\right)^\alpha \leqslant \frac{\omega(y)}{\omega(x)} \leqslant c_2\left(\frac{y}{x}\right)^\beta, \quad 0<x\leqslant y, \quad 0<\alpha\leqslant\beta<1 \tag{3}$$

numbers q, Q, p being such that $1 \leq q \leq \infty$, $Q \geq \frac{\alpha+\beta}{\alpha}$, $\infty \geq p > \frac{Q}{\alpha Q-\alpha-\beta}$, if $p=\infty$ we put $Q=1$.

Let ℓ be a nonnegative function such that

$$\ell^{Q} \in A_1(\partial \mathbb{D}), \tag{4}$$

where A_1 is the Muckenhoupt class [30].

Under such assumptions all closed ideals of $X^{r}_{pq}(\omega, \ell)$ are standard.

As it has been mentioned above in the special case $p=q=\infty$, $\ell = 1$ we obtain the space λ^{r}_{ω} where all closed ideals are also standard under the assumption of only left inequality in (3) [27].

§ 1. An equivalent norm in $X^{r}_{pq}(\omega,\ell)$.

First we introduce equivalent definitions of $X^{r}_{pq}(\omega,\ell)$. Let

$$\Delta^{r+1} f(e^{i\theta}; h) \overset{\text{def}}{=\!=} \sum_{\nu=0}^{r+1} (-1)^{\nu} C^{\nu}_{r+1} f(e^{i(\theta+\nu h)}),$$

we put

$$_s|f|_X \overset{\text{def}}{=\!=} \left\{ \int_0^1 \frac{dh}{h^{1+rq}} \left[\int_0^{2\pi} \left| \frac{\Delta^{r+1} f(e^{i\theta}; h)}{\lambda(\theta; h)} \right|^p d\theta \right]^{q/p} \right\}^{1/q},$$

$$_s\|f\|_X \overset{\text{def}}{=\!=} {}_s|f|_X + \|P_r(f; 0; \cdot)\|_{C_A}.$$

The finiteness of ${}_s\|f\|_X$ or $\|f\|_X$ permits us to write down a formula of the E.M.Dyn'kin type for a function f. Namely, using literally the same arguments as in [37] we obtain the following statement

LEMMA 1.1. Let p, q, Q, ω, ℓ satisfy the conditions of Theorem 14, $f \in C_A$, ${}_s\|f\|_X < \infty$ or $\|f\|_X < \infty$ where $X = X^{r}_{pq}(\omega, \ell)$. Then there exists $v = v_f$ such that

$$\mathcal{J}(v) \overset{def}{=} \left\{ \int_0^1 \frac{dh}{h^{1+(r-1)q}} \left[\int_0^{2\pi} \left| \frac{v((1+h)e^{i\theta})}{\lambda(\theta; h)} \right|^p d\theta \right]^{q/p} \right\}^{1/q} \leqslant C_X \min({}_s\|f\|_X, \|f\|_X) \tag{1.1}$$

and such that

$$f(z) = \iint_{1<|\zeta|<2} \frac{v_f(\zeta)}{\zeta - z} d\sigma_\zeta, \quad z \in \mathbb{D}. \quad \bullet \tag{1.2}$$

DEFINITION 2. Let $f \in C_A$. We put

$$\overset{*}{\Delta}_f(z; h) = \sup_{0 \leqslant r \leqslant 1} \; \sup_{\zeta_1, \zeta_2 \in \overline{\mathbb{D}} \cap D(rz, h)} |f(\zeta_1) - f(\zeta_2)|, \quad z \in \partial\mathbb{D},$$

$$ {}_*|f|_X \overset{def}{=} \left\{ \int_0^1 \frac{dh}{h} \left[\int_0^{2\pi} \left(\frac{\overset{*}{\Delta}_{f^{(r)}}(e^{i\theta}; h)}{\lambda(\theta; h)} \right)^p d\theta \right]^{q/p} \right\}^{1/q},$$

$$ {}_*\|f\|_X \overset{def}{=} {}_*|f|_X + \|P_r(f; 0; \cdot)\|_{C_A}, \quad X = X^{r}_{pq}(\omega, \ell). \tag{1.3}$$

It is clear that $\|f\|_X \leqslant {}_*\|f\|_X$, $X = X^{r}_{pq}(\omega, \ell)$. The main statement of § 1 is the following result.

LEMMA 1.2. Let p, q, Q, ω, ℓ being as in Theorem 14 let $\mathcal{J}(v) < \infty$ for a function v and let f is defined by v in (1.2). Then

$$_{*}|f|_X \leq C_X \mathcal{J}(v), \qquad X = X^{r}_{pq}(\omega, \ell). \tag{1.4}$$

Assuming Lemma 1.2 holds we deduce a corollary which is the main fool in § 2.

COROLLARY. Let p, q, Q, ω, ℓ be as in Theorem 14 $X = X^{r}_{pq}(\omega, \ell)$. Then norms $\| \ \|_X$, $_s\| \ \|_X$, $_*\| \ \|_X$ are equivalent.

PROOF. We can assume $\min(\|f\|_X, \ _s\|f\|_X) < \infty$. Suppose first that $f \in C_A$, $P_r(f; 0; \cdot) \equiv 0$. Then $\|f\|_X = |f|_X$, $_s\|f\|_X = {}_s|f|_X$, $_*\|f\|_X = {}_*|f|_X$. The definition of $| \ |_X$ and $_*| \ |_X$ yields

$$|f|_X \leq {}_*|f|_X, \quad _s|f|_X \leq C_{0X} \, {}_*|f|_X. \tag{1.5}$$

We take $v = v_f$ from Lemma 1.1 such that

$$\mathcal{J}(v) \leq C_X \min(\|f\|_X, \ _s\|f\|_X) = C_X \min(|f|_X, \ _s|f|_X). \tag{1.6}$$

It is possible. Lemma 1.2 and (1.5) imply

$$\max(_s|f|_X, |f|_X) \leq (C_{0X}+1) \cdot {}_*|f|_X \leq C_{1X} \mathcal{J}(v) \leq C_{2X} \min(_s|f|_X, |f|_X). \tag{1.7}$$

In the general case we put $f_0 = f - P_r(f; 0; \cdot)$. Then

$$_s|f|_X = {}_s|f_0|_X, \quad |f|_X = |f_0|_X, \quad _*|f|_X = {}_*|f_0|_X$$

and (1.7) gives

$$\|f\|_X \asymp_* \|f\|_X \asymp_s \|f\|_X . \quad \bullet$$

We proceed to the proof of Lemma 1.2. We notice that for the function ω_0 defined by means of the equality

$$\omega_0(x) = \frac{1}{x}\int_{ax}^{bx} \omega(t)\,dt$$

for suitable a and b depending only on α and β from (3) we have the relations

$$\omega_0(x) \asymp \omega(x) \tag{1.8}$$

and

$$\omega_0'(x) \asymp \frac{\omega_0(x)}{x} . \tag{1.9}$$

So $X^r_{pq}(\omega, \ell) = X^r_{pq}(\omega_0, \ell)$ and the corresponding norms are equivalent.

1.1. Lemmas on the weight $1/\lambda^p$.

LEMMA 1.3. Let $\varkappa > 0$ be a nonnegative function, $\ell^\varkappa \in A_1(\partial\mathbb{D})$. Then for any arc $\gamma \subset I \subset \partial\mathbb{D}$

$$\frac{\int_I \ell(\zeta)|d\zeta|}{\int_\gamma \ell(\zeta)|d\zeta|} \geqslant c\left(\frac{|I|}{|\gamma|}\right)^{1-\frac{1}{\varkappa}}, \quad c = c(\ell). \tag{1.1.1}$$

Proof is an immediate consequence of Lemma 3 from [30] applied to $\ell^\varkappa$. $\bullet$

LEMMA 1.4. Let $f \in L^p(\partial\mathbb{D})$, $p > r > 1$, $\nu = p/r$ and let $w \in A_\nu$ [30]

$$(M_r f(x))^r \stackrel{def}{=} \sup_{I \ni x} \frac{1}{|I|} \int_I |f|^r |d\zeta|.$$

Then

$$\int_{\partial\mathbb{D}} (M_r f)^p w |d\zeta| \leqslant c \int_{\partial\mathbb{D}} |f|^p w |d\zeta|, \quad c = c(p, r, w). \tag{1.1.2}$$

Proof see in [3C] . ●

LEMMA 1.5. Let in the assumptions of Theorem 14 $s = Q/(Q-1)\alpha$, $\nu = p/s$. Then $\lambda^{-p}(\cdot\,; h) \in A_\nu$, A_ν is the Muckenhoupt class and the constant in the definition of A_ν for family of functions $\lambda^{-p}(\cdot\,; h)$ can be chosen not depending on h.

PROOF. The definition of the class A_ν requires the fulfillment of the relation

$$\int_I \lambda^{-p}(x; h)\, dx \cdot \Big(\int_I \lambda^{\frac{p}{\nu-1}}(x; h)\, dx \Big)^{\nu-1} \leqslant c_0 |I|^\nu \tag{1.1.3}$$

with a constant c_0 not depending on the arc $I \subset \partial\mathbb{D}$ and h, $0 < h \leqslant 1$. We put

$$m_x = \operatorname*{ess\,inf}_{\substack{\zeta = e^{i\theta} \\ x \leqslant \theta \leqslant x+h}} \ell(\zeta).$$

Then due to $\ell \in A_1$, we have

$$\lambda(x; h) = \omega\Big(\int_x^{x+h} \ell(e^{i\theta})\, d\theta \Big) \asymp \omega(h m_x).$$

Because $\operatorname*{ess\,inf}_{\theta \leqslant \arg\zeta \leqslant \theta+h} \ell(\zeta) \asymp \operatorname*{ess\,inf}_{\theta \leqslant \arg\zeta \leqslant \theta+2h} \ell(\zeta)$ for $|I| \leqslant h$, $x \in I$ we have $\lambda(x; h) \asymp (h m_{x_0})$, x_0 is the middle of I what gives (1.1.5) in that case. Assume that $h < |I|$ and let

$$\nu(t)=mes\{x\in I: hm_x \geq t\}, \quad m=\operatorname*{ess\,inf}_{x\in I} m_x .$$

We obtain by means of (1.9)

$$\int_I \lambda^{\frac{p}{\nu-1}}(x;h)\,dx \leq c_1 \int_{c_2 hm}^{\infty} \frac{\omega^{\frac{p}{\nu-1}}(t)}{t}\,\nu(t)\,dt. \tag{1.1.4}$$

We notice that $\ell^{\theta} \in A_1$ implies $\ell^{\theta+\delta} \in A_1$ for $\delta>0$ [30] Lemma 5, and in view of Lemma 3 from [30] we get

$$\nu(t) \leq c\left(\frac{hm}{t}\right)^{Q+\delta} |I|. \tag{1.1.5}$$

Inserting (1.1.5) into (1,1,4) and taking into account (3) and the inequality for the number Q, we find

$$\int_I \omega^{\frac{p}{\nu-1}}(hm_x)\,dx \leq c\,\omega^{\frac{p}{\nu-1}}(hm)\cdot|I|. \tag{1.1.6}$$

But it is also clear that

$$\int_I \omega^{-p}(hm_x)\,dx \leq c\,\omega^{-p}(hm)\cdot|I|. \tag{1.1.7}$$

Now (1.1.6) and (1.1.7) prove the lemma. ●

LEMMA 1.6. If $1 \leq q \leq \infty$, $r \geq 0$, $0<\delta_j<1$, $j=1,2$ then for $F \geq 0$ we have

$$\left[\int_0^1 \frac{dh}{h}\left(\int_0^1 \left(\frac{h}{h+\rho}\right)^{\delta_1}\left(\frac{\rho}{h+\rho}\right)^{\delta_2} \frac{F(\rho)}{(h+\rho)^{r}}\,d\rho\right)^{q}\right]^{1/q} \leq c_{\delta_1,\delta_2}\left(\int_0^1 \frac{F^{q}(h)\,dh}{h^{1+(r-1)q}}\right), \tag{1.1.8}$$

$$\Big[\int_0^1 \frac{dh}{h}\Big(\int_0^h \Big(\frac{\rho}{h}\Big)^{\delta_1} \frac{F(\rho)}{\rho^{\tau}}\,d\rho\Big)^q\Big]^{1/q} \leqslant c_{\delta_1}\Big(\int_0^1 \frac{F^q(h)\,dh}{h^{1+(\tau-1)q}}\Big)^{1/q}. \tag{1.1.9}$$

PROOF. The estimates (1.1.8) and (1.1.9) are easy to verify for $q=1$ and $q=\infty$ and we can obtain them by the interpolation theorem of Calderon [38] . ●

LEMMA 1.7. To $0<\rho\leqslant h$ we have

$$c_1\Big(\frac{\rho}{h}\Big)^{\delta_0} \leqslant \frac{\lambda(x;\rho)}{\lambda(x;h)} \leqslant c_2\Big(\frac{\rho}{h}\Big)^{(1-\frac{1}{Q})\alpha} \tag{1.1.10}$$

with a $\delta_0>0$.

PROOF. We only prove the right hand inequality in (1.1.10) because the proof of the left hand inequality is simpler. Then (3) and (1.1.1) imply

$$\frac{\lambda(x;\rho)}{\lambda(x;h)} = \frac{\omega\Big(\int_x^{x+\rho}\ell\Big)}{\omega\Big(\int_x^{x+h}\ell\Big)} \leqslant c\left[\frac{\int_x^{x+\rho}\ell}{\int_x^{x+h}\ell}\right]^{\alpha} \leqslant c\Big(\frac{\rho}{h}\Big)^{(1-\frac{1}{Q})\alpha} .\ ●$$

1.2. PROOF OF LEMMA 1.2.

We finish the proof of Lemma 1.2. Let $\zeta\in\partial\mathbb{D}$, $B\overset{def}{=}D(\zeta,2h)$, $\gamma=B\cap\partial\mathbb{D}$ and let $z_1,z_2\in D(R\zeta,h)$, $0\leqslant R\leqslant 1$.

We have the relation

$$f^{(\tau)}(z_1)-f^{(\tau)}(z_2) = c_\tau\iint_B \frac{v(\xi)}{(\xi-z_1)^{\tau+1}}\,d\sigma - c_\tau\iint_B \frac{v(\xi)}{(\xi-z_2)^{\tau+1}}\,d\sigma +$$

$$+ (z_2-z_1)\,c_\tau\iint_B \frac{v(\xi)\big[(\xi-z_1)^{\tau}+(\xi-z_1)^{\tau-1}(\xi-z_2)+\ldots+(\xi-z_2)^{\tau}\big]}{(\xi-z_1)^{\tau+1}(\xi-z_2)^{\tau+1}}\,d\sigma. \tag{1.2.1}$$

Let $z_1^o, z_2^o \in \gamma$ be the following points

$$z_j^o = \begin{cases} \dfrac{z_j + (1-R)\zeta}{|z_j + (1-R)\zeta|}, & \text{if } |z_j + (1-R)\zeta| \leqslant 1, \\ z_j + \tau\zeta, \quad 0 \leqslant \tau < 1-R, & \text{if } |z_j + (1-R)\zeta| > 1, \end{cases}$$

$j = 1, 2.$

Then

$$\left| \iint\limits_B \frac{v(\xi)}{(\xi - z_1)^{r+1}} \, d\sigma \right| \leqslant c \iint\limits_B \frac{|v(\xi)|}{|\xi - z_1^o|^{r+1}} \, d\sigma,$$

and that estimate holds for two other integrals in (1.2.1). Finally we get

$$|f^{(r)}(z_1) - f^{(r)}(z_1)| \leqslant c \Big[\sup_{z \in \gamma} \iint\limits_B \frac{|v(\xi)|}{|\xi - z|^{r+1}} \, d\sigma + h \iint\limits_{C \setminus B} \frac{|v(\xi)|}{|\xi - \zeta|^{r+2}} \, d\sigma \Big] \leqslant$$

$$\leqslant c \Big[\sup_{z \in \gamma} \iint\limits_B \frac{|v(\xi)|}{|\xi - z|^{r+1}} \, d\sigma + \int_0^1 \frac{h}{(h+\rho)^{r+1}} \, Mv((1+\rho)\zeta) \, d\rho \Big],$$

where $Mv((1+\rho)\zeta) \overset{\mathrm{def}}{=} M_1(v((1+\rho)\zeta))$ and for $1 \leqslant r < \infty$

$$M_\tau^r(v)((1+\rho)\zeta) = \sup_{0 < |\lambda| \leqslant \pi} \frac{1}{|\lambda|} \left| \int_0^\lambda |v((1+\rho)\zeta e^{id})|^r \, dr \right|.$$

Now with the help of Lemma 1.5 and Muckenhoupt Theorem we obtain

$$\int_0^1 \frac{dh}{h} \Big[\int_0^{2\pi} \frac{dx}{\lambda^p(x;h)} \Big(\int_0^1 \frac{h}{(h+\rho)^{r+1}} \, Mv((1+\rho)e^{ix}) \, d\rho \Big)^p \Big]^{q/p} \leqslant$$

$$\leqslant \int_0^1 \frac{dh}{h} \Big[\int_0^1 \frac{h}{(h+\rho)^{r+1}} \, d\rho \Big(\int_0^{2\pi} \Big(\frac{Mv((1+\rho)e^{ix})}{\lambda(x;h)} \Big)^p dx \Big)^{1/p} \Big]^q \leqslant$$

$$\leq c\int_0^1 \frac{dh}{h}\Big[\int_0^1 \frac{h}{(h+\rho)^{\tau+1}}\, d\rho\Big(\int_0^{2\pi}\Big|\frac{v((1+\rho)e^{ix})}{\lambda(x;h)}\Big|^p dx\Big)^{1/p}\Big]^q \leq$$

$$\leq c\int_0^1 \frac{dh}{h}\Big[\int_0^1 \frac{h}{(h+\rho)^{\tau+1}}\cdot\Big(\frac{h+\rho}{h}\Big)^{1-\delta_0}\cdot\Big(\frac{\rho}{h+\rho}\Big)^{\delta_0} d\rho\Big(\int_0^{2\pi}\Big|\frac{v((1+\rho)e^{ix}}{\lambda(x;\rho)}\Big|^p dx\Big)^{1/p}\Big]^q \leq$$

$$\leq c\int_0^1 \frac{dh}{h^{1+(\tau-1)q}}\Big(\int_0^{2\pi}\Big|\frac{v((1+\rho)e^{ix})}{\lambda(x;\rho)}\Big|^p dx\Big)^{q/p}. \tag{1.2.2}$$

We study now the contribution of $\iint\limits_B$. We put $t = Q/(Q-1)\alpha$, t' is such that $1/t + 1/t' = 1$, $\zeta = e^{i\theta_0}$, $z_0 \in \gamma$. We have

$$\iint\limits_B \Big|\frac{v(\xi)}{(\xi - z_0)^{\tau+1}}\Big| d\sigma_\xi \leq 2\int_0^h d\rho\Big(\int_{\theta_0-h}^{\theta_0+h} |v((1+\rho)e^{i\theta})|^t d\theta\Big)^{1/t}\cdot\Big(\int_{\theta_0-h}^{\theta_0+h}\frac{d\theta}{|(1+\rho)e^{i\theta}-z_0|^{t'(\tau+1)}}\Big)^{1/t'} \leq$$

$$\leq c\int_0^h M_t\,(v((1+\rho)\zeta)\cdot\Big(\frac{h}{\rho}\Big)^{1/t}\frac{d\rho}{\rho^\tau}. \tag{1.2.3}$$

But $\ell^Q \in A_1$ implies $\ell^{Q_1} \in A_1$ for a $Q_1 > Q$ [30] and then at once in view of Lemmas 1.5 - 1.7, the Muckenhoupt theorem and (1.2.3), we obtain

$$\int_0^1 \frac{dh}{h}\Big[\int_0^{2\pi}\frac{dx}{\lambda^p(x;h)}\Big|\iint\limits_B\Big|^p\Big]^{q/p} \leq$$

$$\leq c\int_0^1 \frac{dh}{h}\Big[\int_0^{2\pi}\frac{dx}{\lambda^p(x;h)}\Big(\int_0^h M_t(v((1+\rho)e^{ix}))\Big(\frac{h}{\rho}\Big)^{1/t}\frac{d\rho}{\rho^\tau}\Big)^p\Big]^{q/p} \leq$$

$$\leq c\int_0^1 \frac{dh}{h}\Big[\int_0^h\Big(\frac{h}{\rho}\Big)^{1/t}\frac{d\rho}{\rho^\tau}\Big(\int_0^{2\pi}\Big(\frac{M_t(v)((1+\rho)e^{ix}}{\lambda(x;h)}\Big)^p dx\Big)^{1/p}\Big]^q \leq$$

$$\leqslant c\int_0^1 \frac{dh}{h}\Big[\int_0^h \Big(\frac{h}{\rho}\Big)^{1/t}\frac{d\rho}{\rho^{r}}\Big(\int_0^{2\pi}\Big|\frac{v((1+\rho)e^{ix}}{\lambda(x;h)}\Big|^p dx\Big)^{1/p}\Big]^q \leqslant$$

$$\leqslant c\int_0^1 \frac{dh}{h}\Big[\int_0^h \Big(\frac{\rho}{h}\Big)^{(1-\frac{1}{Q_1})\alpha-\frac{1}{t}}\frac{d\rho}{\rho^{r}}\Big(\int_0^{2\pi}\Big|\frac{v((1+\rho)e^{ix}}{\lambda(x;\rho)}\Big|^p dx\Big)^{1/p}\Big]^q \leqslant$$

$$\leqslant c\int_0^1 \frac{dh}{h^{1+(r-1)q}}\Big(\int_0^{2\pi}\Big|\frac{v((1+\rho)e^{ix}}{\lambda(x;\rho)}\Big|^p dx\Big)^{q/p},$$

(1.2.4)

Since $(1-\frac{1}{Q_1})\alpha > \frac{1}{t}$ and $\lambda^{-p}(\cdot\,;h)\in A_\nu$, $\nu = p/t$. The relations (1.2.2) and (1.2.4) prove Lemma 1.2. ●

§ 2. A special approximation in $X^r_{pq}(\omega,\ell)$.

In this section we finish the proof of Theorem 14. Arguing exactly as in [23] - [27] it is easy to check that only the approximation problem discussed in the Introduction requires additional effects . Moreover, according to an idea of B.I.Korenblum [23] , [24] one can only prove the following results.

STATEMENT 1. Let $f\in X^r_{pq}(\omega,\ell)$, $E=\bigcap_{\nu=0}^{r} Z_{f^{(\nu)}}(\bar{\mathbb{D}})\cap\partial\mathbb{D}$, $d=d_E$ be an outer function in $\mathbb{D}$ such that

$$|d(e^{i\theta})| = \frac{(\beta-\theta)(\theta-\alpha)}{\beta-\alpha},$$

for $\theta\in(\alpha,\beta)\subset[0,2\pi)\setminus\arg E$, $\alpha,\beta\in\arg E$.

Then for $0 < \varepsilon < 1$ we have

$$f d^{\varepsilon} \in X^{r}_{pq}(\omega, \ell)$$

and moreover

$$\|f d^{\varepsilon}\|_X \leq c_0 \|f\|_X, \quad X = X^{r}_{pq}(\omega, \ell)$$

where C_0 does not depend on ε.

STATEMENT 2. Keeping notations of Statement 1 we put $f_0 = f d^{\varepsilon}$ and let τ be an arbitrary set of complementary arcs of the set E, F_{τ} be an outer function in $\mathbb{D}$ defined by the following relation

$$|F_{\tau}(z)| = \begin{cases} |f_0(z)|, & z \in \tau, \\ 1, & z \in \partial\mathbb{D} \setminus \tau, \end{cases}$$

N be a fix natural number,

$$\varphi = f_0 F_{\tau}^{N}.$$

Then $\varphi \in X^{r}_{pq}(\omega, \ell)$ and moreover

$$\|\varphi\|_X \leq C_1 \|f\|_X,$$

where the constant C_1 does not depend on τ.

The proof of the statements requires some lemmas.

2.1. Auxiliary lemmas.

LEMMA 2.1. We have $X^{r}_{pq}(\omega, \ell) \subset \Lambda^{r+\nu}$, $\nu = \alpha(1 - \frac{1}{Q}) - \frac{1}{p}$.

PROOF. The inclusion is to check only for $r = 0$. The corollary of Lemma 1.2 implies that for $f \in X^{0}_{pq}(\omega, \ell)$ we have

$$\int_{2H}^{4H} \frac{dh}{h}\left(\int_{\theta}^{\theta+H} \left| \frac{\Delta_f^*(x;2h)}{\lambda(x;2h)}\right|^p dx\right)^{q/p} \leqslant c,$$

Hence by Lemma 1.7

$$|f(e^{i(\theta+H)}) - f(e^{i\theta})| \leqslant C\,\frac{\lambda(x;H)}{H^{1/p}} \leqslant C_1 H^{\alpha(1-\frac{1}{Q})-\frac{1}{p}}. \quad \bullet$$

LEMMA 2.2. Let $f \in C^{\nu}[0,\infty)$, $\nu \geqslant 0$, $f^{(\mu)}(0) = 0$, $0 \leqslant \mu \leqslant \nu$.

We put

$$f^+(x) = \max_{\substack{t \geqslant 0 \\ t\in[x-c_0h,\,x+c_0h]}} |f| \quad , \quad c_0 > 1$$

$$\Delta^+_{f^{(\nu)}}(x;h) = \max_{\substack{y_1,y_2 \in [x-h,\,x+h] \\ y_j \geqslant 0}} |f^{(\nu)}(y_1) - f^{(\nu)}(y_2)|$$

and let $1 < p < \infty$, $w \in A_p(\mathbb{R}^+)$. Then

$$\int_h^{\infty} \left(\frac{f^+(x)}{x^{\nu+1}}\right)^p w(x)\,dx \leqslant c \int_0^{\infty} \left(\frac{\Delta^+_{f^{(\nu)}}(x;2h)}{2h}\right)^p w(x)\,dx,$$

where c depends only on p, ν, w and c_0.

PROOF. Suppose first $\nu \geqslant 1$. If $x \geqslant h$ and $x_0 \geqslant 0$ is such a number that $|f(x_0)| = f^+(x)$, $|x - x_0| \leqslant c_0 h$ then

$$|f(x_0)| \underset{\nu}{\asymp} \left|\int_0^{x_0} (x_0-t)^{\nu-1} f^{(\nu)}(t)\,dt\right| \underset{\nu,c_0}{\leqslant} x^{\nu-1} \int_0^{x+c_0h} |f^{(\nu)}(t)|\,dt \leqslant$$

$$\leqslant x^{\nu-1}(x+h) \int_0^{(1+c_0)x} \frac{|f^{(\nu)}(t)|}{t+h}\,dt \underset{c_0}{\leqslant} x^{\nu+1} M\varphi_f(x),$$

$$\varphi_f(t) = \frac{f^{(\nu)}(t)}{t+h},$$

and then the Muckenhoupt theorem yields

$$\int_h^\infty \left(\frac{f^+(x)}{x^{\nu+1}}\right)^p w(x)dx \underset{\nu,c_0}{\lessdot} \int_h^\infty (M\varphi_f(x))^p w(x)\,dx \underset{\nu,c_0,w,p}{\lessdot}$$

$$\underset{\nu,c_0,w,p}{\lessdot} \int_0^\infty \left|\frac{f^{(\nu)}(t)}{t+h}\right|^p w(t)dt \lessdot \int_h^\infty \left|\frac{f^{(\nu)}(t)}{t}\right|^p w(t)\,dt +$$

$$+\int_0^\infty \left(\frac{\Delta^+_{f^{(\nu)}}(x;2h)}{2h}\right)^p w(x)\,dx, \tag{2.1.1}$$

because

$$|f^{(\nu)}(t)| = |f^{(\nu)}(t) - f^{(\nu)}(0)| \leqslant \Delta^+_{f^{(\nu)}}(t;2h), \quad t \leqslant h.$$

The relation (2.1.1) permits us to consider only the case $\nu = 0$. We define further the function $f^\times(x) = \max\limits_{[(2n-1)h,(2n+1)h]} |f|$ if $x \in [(2n-1)h,(2n+1)h)$, $n \geqslant 1$,

Then taking into account that for two arcs $J \supset I$, $|J| \leqslant |I|$,

$$\int_J w \lessdot \int_I w$$

we have

$$\int_h^\infty \left(\frac{f^+(x)}{x}\right)^p w(x)\,dx \underset{c_0,w}{\lessdot} \int_h^\infty \left(\frac{f^\times(x)}{x}\right)^p w(x)\,dx. \tag{2.1.2}$$

Let $f^\times(x) = |f(x_n)|$ where $x \in [(2n-1)h,(2n+1)h)$, $x_n \in [(2n-1)h,(2n+1)h]$.

Then

$$\int_{(2n-1)h}^{(2n+1)h} \left(\frac{f^{x}(x)}{x}\right)^{p} w(x)\,dx \leq \int_{(2n-1)h}^{(2n+1)h} \left|\frac{f^{x}(x)-|f(x)|}{x}\right|^{p} w(x)\,dx +$$

$$+ \int_{(2n-1)h}^{(2n+1)h} \left|\frac{f(x)}{x}\right|^{p} w(x)\,dx = \int_{(2n-1)h}^{(2n+1)h} \left|\frac{|f(x_n)|-|f(x)|}{x}\right|^{p} w(x)\,dx +$$

$$+ \int_{(2n-1)h}^{(2n+1)h} \left|\frac{f(x)}{x}\right|^{p} w(x)\,dx \leq \int_{(2n-1)h}^{(2n+1)h} \left(\frac{\Delta_f^{+}(x;2h)}{2h}\right)^{p} w(x)\,dx +$$

$$+ \int_{(2n-1)h}^{(2n+1)h} \left|\frac{f(x)}{x}\right|^{p} w(x)\,dx, \qquad (2.1.3)$$

and hence (2.1.2) and (2.1.3) yield

$$\int_{h}^{\infty} \left(\frac{f^{+}(x)}{x}\right)^{p} w(x)\,dx \leq \int_{h}^{\infty} \left|\frac{f(x)}{x}\right|^{p} w(x)\,dx + \int_{0}^{\infty} \left(\frac{\Delta_f^{+}(x;2h)}{2h}\right)^{p} w(x)\,dx.$$

But now if $f_0 \in C[0,\infty)$, $f_0(0)=0$, $f_0' \in L_w^p(0,\infty)$ then due to

$$\left|\frac{f_0(x)}{x}\right| = \left|\frac{1}{x}\int_0^x f_0'(t)\,dt\right| \leq M f_0'(x),$$

again with the help of Muckenhoupt theorem we get

$$\int_0^\infty \left|\frac{f_0(x)}{x}\right|^{p} w(x)\,dx \underset{p,w}{\leq} \int_0^\infty |f_0'(x)|^{p} w(x)\,dx. \qquad (2.1.4)$$

Assume now that $f \in C[0,\infty)$, $f(0)=0$. We extend the function on $(-\infty, 0)$ as zero, the weight w we extend to be an even function and put

$$f_0(x)=\frac{1}{2h}\int_{x-h}^{x+h} f(t)\,dt.$$

Then $f_0(-h)=0, \quad f_0'(x)=\frac{1}{2h}(f(x+h)-f(x-h))$. Hence due to (2.1.4) we obtain

$$\int_h^\infty\left|\frac{f_0(x)}{x}\right|^p w(x)\,dx\leqslant\int_{-h}^\infty|f_0'(x)|^p w(x)\,dx\leqslant$$

$$\leqslant\int_h^\infty\left|\frac{f(x+h)-f(x-h)}{2h}\right|^p w(x)\,dx+\int_0^h\left(\frac{|f(h-x)|+|f(h+x)|}{2h}\right)^p w(x)\,dx\leqslant$$

$$\leqslant\int_0^\infty\left(\frac{\Delta_f^+(x;2h)}{2h}\right)^p w(x)\,dx. \tag{2.1.5}$$

Next for $x\geqslant h$ we have

$$\left|\frac{f(x)-f_0(x)}{x}\right|=\frac{1}{2hx}\left|\int_{x-h}^{x+h}(f(x)-f(t))\,dt\right|\leqslant$$

$$\leqslant\frac{1}{x}\,\Delta_f^+(x;h)\leqslant\frac{1}{h}\,\Delta_f^+(x;2h). \tag{2.1.6}$$

The relations (2.1.5) and (2.1.6) proves the Lemma. ●

LEMMA 2.3. We have

$$\int_0^1\frac{dh}{h^{1-q}}\left(\int_0^{2\pi}\frac{dx}{\lambda^p(x;h)}\right)^{q/p}<\infty. \tag{2.1.7}$$

PROOF. Because $l\in A_1$ we have $\int_x^{x+h} l\geqslant mh$. The assumption gives $\omega(y)\geqslant cy^\beta$, i.e. $\lambda(x;h)\geqslant ch, \quad 0<h<1$ and then

$$(2.17) \leq c_1 \int_0^1 \frac{dh}{h^{1-q(1-\beta)}} < \infty. \quad \bullet$$

LEMMA 2.4. Let $d = d_E$ be an outer function from Statement 1, $\rho(z) = dist(z, E)$ the function $a(z)$ be defined for $z \in \partial\mathbb{D} \setminus E$ as follows:

$$a(z) = \int_{\partial\mathbb{D} \setminus \gamma} \frac{|\log|d(\zeta)||}{|\zeta - z|^2} |d\zeta|,$$

for $z \in \gamma$, γ be any complementary arc of E. Then for $0 < \varepsilon < 1$ we have

$$|(d^\varepsilon(z))^{(\nu)}| \leq C_\nu (\varepsilon^\nu a^\nu(z) \rho^\varepsilon(z) + \rho^{\frac{\varepsilon}{2} - \nu}(z)), \quad z \in \partial\mathbb{D},$$

and the constant C_ν does not depend on z and ε.

PROOF. Let $v = \log d$. Then [23]

$$|v^{(r)}(z)| \lesssim_r \frac{a(z)}{\rho^{r-1}(z)} + \frac{\log \frac{4}{\rho(z)}}{\rho^r(z)}, \quad r \geq 1.$$

Next $d^\varepsilon = e^{\varepsilon v}$ hence

$$(d^\varepsilon)^{(\nu)} = d^\varepsilon \sum_{(r,\mu)} A^{r_1,\ldots,r_\ell}_{\mu_1,\ldots,\mu_\ell} \prod_{j=1}^{\ell} (\varepsilon v^{(r_j)})^{\mu_j},$$

where $\sum_{j=1}^{\ell} r_j \mu_j = \nu$. Hence

$$|(d^\varepsilon)^{(\nu)}| \lesssim_\nu \rho^\varepsilon \sum_{(r,\mu)} \prod_{j=1}^{\ell} \left(\frac{\varepsilon a}{\rho^{r_j - 1}} + \frac{\varepsilon \log \frac{4}{\rho}}{\rho^{r_j}} \right)^{\mu_j} = \rho^{\varepsilon - \nu} \sum_{(r,\mu)} (a\varepsilon\rho + \varepsilon \log \frac{4}{\rho})^{\mu_0} \qquad (2.1.8)$$

where $\mu_0 = \sum_{j=1}^{\ell} \mu_j \leq \nu$ and

$$(2.1.8) \underset{\nu}{\lesssim} \rho^{\varepsilon-\nu} \sum_{(r,\mu)} (a\varepsilon\rho)^{\mu_0} + \rho^{\varepsilon-\nu} \sum_{(r,\mu)} \varepsilon^{\mu_0} (\log \tfrac{4}{\rho})^{\mu_0}. \tag{2.1.9}$$

We observe that $\varepsilon^{\mu_0}\rho^{\varepsilon}(\log\frac{4}{\rho})^{\mu_0} \underset{\nu}{\lesssim} \rho^{\varepsilon/2}$ due to $\mu_0 \geq 1$. If $\rho a\varepsilon < 1$ then

$$\rho^{\varepsilon-\nu} \sum_{(r,\mu)} (\rho a\varepsilon)^{\mu_0} \underset{\nu}{\lesssim} \rho^{\varepsilon-\nu},$$

but if $\rho a\varepsilon \geq 1$ then $(\rho a\varepsilon)^{\mu_0} \leq (\rho a\varepsilon)^{\nu}$ and in any case we have

$$(2.1.9) \underset{\nu}{\lesssim} \varepsilon^{\nu} a^{\nu} \rho^{\varepsilon} + \rho^{\frac{\varepsilon}{2}-\nu}. \qquad \bullet$$

COROLLARY. If $a(z) \leq A \dfrac{\log\frac{4}{\rho(z)}}{\rho(z)}$ then

$$|(d^{\varepsilon})^{(\nu)}(z)| \underset{\nu,A}{\lesssim} \rho^{-\nu}(z). \qquad \bullet$$

LEMMA 2.5. Let $f \in C_A$. We define

$$a_f(z) = \int_{\partial\mathbb{D}\setminus\gamma} \frac{|\log|f(\zeta)||}{|\zeta - z|^2}\,|d\zeta|,$$

for $z \in \gamma$, γ is a complementary arc of $Z_f(\partial\mathbb{D})$ and let $\mu_f(z) = a_f(z)/\log\frac{4}{\rho(z)}$. There exists a constant $A = A(r)$ such that if $\mu_f(z) > A\rho^{-1}(z)$ then for $\zeta \in \partial\mathbb{D}$, $|\zeta - z| \leq \frac{1}{2}\rho(z)$ holds

$$|f^{(\nu)}((1-\tfrac{1}{\mu_f(z)})\zeta)| \leq c\rho^{2(r+1)}(z), \quad 0 \leq \nu \leq r+1.$$

Proof is a consequence of the integral representation of the outer part of a function (see too [23] , [24]). ●

LEMMA 2.6. If $f \in X^{r}_{pq}(\omega, l)$ be an inner function $f/I \in C_A$ then $f/I \in X^{r}_{pq}(\omega, l)$ and $\|f/I\|_X \leqslant c_0 \|f\|_X$, and the constant c_0 does not depend on f and I.

PROOF. If $f \in X^{r}_{pq}(\omega, l)$ then by Lemma 1.1

$$f(z) = \iint\limits_{1<|\zeta|<2} \frac{v(\zeta)}{\zeta - z}\, d\sigma_\zeta$$

and following E.M.Dyn'kin [39] we obtain from $f/I \in C_A$ that

$$\frac{f(z)}{I(z)} = \iint\limits_{1<|\zeta|<2} \frac{v(\zeta)}{I(\zeta)} \frac{d\sigma}{\zeta - z},$$

where the function v is chosen as in Lemma 1.2. Since for $|\zeta| > 1$ $|I^{-1}(\zeta)| \leqslant 1$ we have $|v(\zeta)/I(\zeta)| \leqslant$ $\leqslant |v(\zeta)|$ and then Lemma 2.6 follows from Lemma 1.2. ●

2.2. PROOF OF STATEMENT 1.

We proceed now to the proof of Statement 1. We fix a function $f \in X^{r}_{pq}(\omega, l)$, $\|f\|_X = 1$ and a number A from Lemma 2.5 and define a decomposition of $\partial\mathbb{D}$ into h, $0<h\leqslant 1$ four disjoint sets $T_j(h)$, $1 \leqslant j \leqslant 4$, $(\rho(z))$ is $dist(z, E)$, E is the set from Statement 1.

$$T_1(h) = \{ z \in \partial\mathbb{D} : \rho(z) \leqslant 2(r+1)h \};$$

$$T_2(h) = \left\{ z \in \partial\mathbb{D} : \rho(z) > 2(r+1)h, \quad a(z) \leqslant AC_0 \frac{\log \frac{4}{\rho(z)}}{\rho(z)} \right\};$$

$$T_3(h) = \left\{ z \in \partial\mathbb{D} : \rho(z) > 2(r+1)h, \quad az > AC_0 \frac{\log \frac{4}{\rho(z)}}{\rho(z)}, \quad h \geqslant \frac{1}{\mu_f(z)} \right\};$$

$$T_4(h) = \left\{ z \in \partial\mathbb{D} : \rho(z) > 2(r+1)h, \quad az > AC_0 \frac{\log \frac{4}{\rho(z)}}{\rho(z)}, \quad h < \frac{1}{\mu_f(z)} \right\},$$

where

$$\mu_f(z) = \frac{a_f(z)}{\log \frac{4}{\rho(z)}}$$

and the number C_0 is chosen such that in the notations of Lemma 2.5 $a(z) \leqslant c_0 a_f(z)$ for $z \in \partial\mathbb{D}$. Such a choice of c_0 is possible due to Lemma 2.1. We shall prove the following result

$$_s\|fd^{\varepsilon}\|_X \leqslant const \cdot {}_*\|f\|_X. \tag{2.2.1}$$

and the constant does not depend on ε. Due to theorem of S.Banach it is sufficiently to verify that

$$\left\{\int_0^1 \frac{dh}{h^{1+rq}}\left[\int_0^{2\pi}\left|\frac{\Delta^{r+1}(d^{\varepsilon}f)(e^{i\theta};h)}{\lambda(\theta;h)}\right|^p d\theta\right]^{q/p}\right\}^{1/q} < \infty$$

and that will be proved if we state four relations

$$\left\{\int_0^1 \frac{dh}{h^{1+rq}}\left[\int_{\arg T_j(h)} |\cdot|^p\, d\theta\right]^{q/p}\right\}^{1/q} < \infty, \qquad j = 1,2,3,4. \tag{2.2.2}$$

$(2.2.2_1)$ because $|d| \leqslant 1$ we have

$$|\Delta^{r+1}(d^{\varepsilon}f)(e^{i\theta};h)| \underset{r}{\leqslant} \sum_{\nu=0}^{r+1} |f(e^{i(\theta+\nu h)})|,$$

and then

$$(2.2.2_1) \leqslant \sum_{\nu=0}^{r+1}\left[\int_0^1 \frac{dh}{h^{1+rq}}\left(\int_{\arg T_1(h)}\left|\frac{f(e^{i(\theta+\nu h)}}{\lambda(\theta;h)}\right|^p d\theta\right)^{q/p}\right]^{1/q} \underset{r}{\leqslant}$$

$$\underset{r}{\preccurlyeq}\left[\int_0^1 \frac{dh}{h^{1+rq}}\left(\int_{\arg T_1^*(h)}\left|\frac{f(e^{i\theta})}{\lambda(\theta;h)}\right|^p d\theta\right)^{q/p}\right]^{1/q}, \tag{2.2.3}$$

where $T_1^*(h)=\{z\in\partial\mathbb{D}:\rho(z)\leqslant\frac{\pi}{2}\cdot 4(r+1)h\}$ because for

$$e^{i\theta}\in T_1(h) \quad e^{i(\theta+\nu h)}\in T_1^*(h),\quad 0\leqslant\nu\leqslant r+1,$$

and

$$\lambda(\theta;h)\underset{r,l}{\asymp}\lambda(\theta+\nu h;h),\quad 0\leqslant\nu\leqslant r+1.$$

Further if $z=e^{i\theta}\in T_1^*(h)$, $z_0\in E$ is the nearest point to z then

$$|f(z)|=\left|\frac{1}{(r-1)!}\int_{z_0}^{z}(z-t)^{r-1}(f^{(r)}(t)-f^{(r)}(z_0))dt\right|\underset{r}{\preccurlyeq}$$

$$\underset{r}{\preccurlyeq}|z-z_0|^r\max_{\substack{|t-z_0|\leqslant|z-z_0|\\ t\in\partial\mathbb{D}}}|f^{(r)}(t)-f^{(r)}(z_0)|\underset{r}{\preccurlyeq}h^r\Delta^*_{f^{(r)}}(e^{i\theta};8(r+1)h).$$

Hence

$$(2.2.3)\preccurlyeq\left[\int_0^1\frac{dh}{h}\left(\int_0^{2\pi}\left(\frac{\Delta^*_{f^{(r)}}(e^{i\theta};8(r+1)h)}{\lambda(\theta;h)}\right)^p d\theta\right)^{q/p}\right]^{1/q}\preccurlyeq$$

$$\preccurlyeq\left[\int_0^1\frac{dh}{h}\left(\int_0^{2\pi}\left(\frac{\Delta^*_{f^{(r)}}(e^{i\theta};h)}{\lambda(\theta;h)}\right)^p d\theta\right)^{q/p}\right]^{1/q}<\infty \tag{2.2.4}$$

by Lemma 1.2 because $\lambda(\theta;h)\asymp\lambda(\theta,\nu h)$, $1\leqslant\nu\leqslant 8(r+1)h$.

(2.2.2$_2$) we use the formula

$$\Delta^r\varphi(x;h)=\underbrace{\int_0^h\cdots\int_0^h}_{r}\varphi^{(r)}(x+t_1+\ldots+t_r)\,dt_1\ldots dt_r,$$

$$\Delta^{r+1}\varphi(x;h) = \Delta^{r}\varphi(x+h;h) - \Delta^{r}\varphi(x;h) =$$

$$= \underbrace{\int_0^h \dots \int_0^h}_{r} (\varphi^{(r)}(x+h+t_1+\dots+t_r) - \varphi^{(r)}(x+t_1+\dots+t_r))\,dt_1\dots dt_r .$$

Let $\partial = \dfrac{\partial}{\partial\theta}$. Then (2.2.3) implies

$$\Delta^{r+1}\varphi(e^{i\theta};h) = \int_0^h \dots \int_0^h (\partial^r \varphi(e^{i(\theta+h+t_1+\dots+t_r)}) - \partial^r \varphi(e^{i(\theta+t_1+\dots+t_r)}))\,dt_1\dots dt_r =$$

$$= \int_0^h \dots \int_0^h (-i)^r \left[e^{ir\theta_1}\varphi^{(r)}(e^{i(\theta+h+t_1+\dots+t_r)}) - e^{ir\theta_2}\varphi^{(r)}(e^{i(\theta+t_1+\dots+t_r)})\right] dt_1\dots dt_r +$$

$$+ \int_0^h \dots \int_0^h (T\varphi(e^{i(\theta+h+t_1+\dots+t_r)}) - T\varphi(e^{i(\theta+t_1+\dots+t_r)}))\,dt_1\dots dt_r ,$$

$$\theta_1 = \theta+h+t_1+\dots+t_r , \quad \theta_2 = \theta+t_1+\dots+t_r , \tag{2.2.6}$$

where

$$T = \sum_{j=0}^{r-1} b_j(\theta)\left(\frac{d}{dz}\right)^j, \quad b_j \in C^\infty(0,2\pi), \quad z = e^{i\theta}.$$

Next, we have the inclusion

$$X^{\nu}_{pq}(\omega,l) \subset X^{r}_{pq}(\omega,l), \qquad 0 \leqslant \nu \leqslant r.$$

Hence every summand in the sum T will be estimated as the main term in the following arguments (2.2.7) - (2.2.14):

$$\left| \int_0^h \dots \int_0^h e^{ir\theta_1} (\varphi^{(r)}(e^{i(\theta+h+t_1+\dots+t_r)}) - \varphi^{(r)}(e^{i(\theta+t_1+\dots+t_r)}))\,dt_1\dots dt_r \right| \leqslant$$

$$\leqslant h^r \max_{0\leqslant\tau\leqslant rh} |\varphi^{(r)}(e^{i(\theta+h+\tau)}) - \varphi^{(r)}(e^{i(\theta+\tau)})|.$$

Let $\varphi = f d^{\varepsilon}$. Then

$$\varphi^{(r)}(z_1) - \varphi^{(r)}(z_2) = \sum_{\nu=0}^{r} c_2^{\nu}([f^{(\nu)}(z_1)(d^{\varepsilon})^{(r-\nu)}(z_1) - f^{(\nu)}(z_2)(d^{\varepsilon})^{(r-\nu)}(z_2)] =$$

$$= \sum_{\nu=0}^{r} c_r^{\nu}(f^{(\nu)}(z_1) - f^{(\nu)}(z_2))(d^{\varepsilon})^{(r-\nu)}(z_1) + \sum_{\nu=0}^{r} c_r^{\nu} f^{(\nu)}(z_2)((d^{\varepsilon})^{(r-\nu)}(z_1) - (d^{\varepsilon})^{(r-\nu)}(z_2)) =$$

$$= \sum\nolimits_1 + \sum\nolimits_2 , \tag{2.2.7}$$

$$e^{ir\theta_1}\varphi^{(r)}(e^{i\theta_1}) - e^{ir\theta_2}\varphi^{(r)}(e^{i\theta_2}) =$$

$$= e^{ir\theta_1}(\varphi^{(r)}(e^{i\theta_1}) - \varphi^{(r)}(e^{i\theta_2})) + \varphi^{(r)}(e^{i\theta_2})(e^{ir\theta_2} - e^{ir\theta_1}) \overset{def}{=}$$

$$\overset{def}{=} e^{ir\theta_1}(\sum\nolimits_1 + \sum\nolimits_2) + \sum\nolimits_3 . \tag{2.2.7'}$$

Next, if $z = e^{i\theta} \in T_2(h)$, $|\tau| \leqslant rh$ then Lemma 2.5 yields

$$|(d^{\varepsilon})^{(r-\nu)}(z)| \preccurlyeq \rho(z)^{\nu-r}, \tag{2.2.8}$$

$$|(d^{\varepsilon})^{(r-\nu)}(e^{i(\theta+\tau+h)}) - (d^{\varepsilon})^{(r-\nu)}(e^{i(\theta+\tau)}) \preccurlyeq h\rho(z)^{\nu-r-1} . \tag{2.2.9}$$

We put for $0 \leqslant \nu \leqslant r$ $f_{\nu}(z) \overset{def}{=} \max_{0 \leqslant \lambda \leqslant (r+1)h} |f^{(\nu)}(e^{i(\theta+\lambda)})|$

then we have

$$|(f d^{\varepsilon})^{(r)}(e^{i(\theta+\tau)})| \preccurlyeq \sum_{\nu=0}^{r} \rho(z)^{\nu-r} \cdot f_{\nu}(z) \tag{2.2.10}$$

and for $0 \leq \nu \leq r-1$ we write

$$|f^{(\nu)}(e^{i(\theta+\tau+h)}) - f^{(\nu)}(e^{i(\theta+\tau)})| \leq h f_{\nu+1}(z), \qquad (2.2.10')$$

$$|f^{(r)}(e^{i(\theta+\tau+h)}) - f^{(r)}(e^{i(\theta+\tau)})| \leq \Delta^{*}_{f^{(r)}}(e^{i\theta}; 2(r+1)h). \qquad (2.2.11)$$

Hence (2.2.7) - (2.2.11) imply

$$\int_0^1 \frac{dh}{h^{1+rq}} \Big[\int_{T_2(h)} \Big| \frac{\Sigma_1}{\lambda(\theta;h)} \Big|^p d\theta \Big]^{q/p} \leq \sum_{\nu=0}^{r-1} \int_0^1 \frac{dh}{h} \Big[\int_{T_2(h)} \Big(\frac{h f_{\nu+1}(z)}{\rho^{r-\nu}(z)\lambda(\theta;h)} \Big)^p d\theta \Big]^{q/p} +$$

$$+ \int_0^1 \frac{dh}{h} \Big[\int_{\partial \mathbb{D}} \Big(\frac{\Delta^{*}_{f^{(r)}}(e^{i\theta}; 2(r+1)h)}{\lambda(\theta;h)} \Big)^p d\theta \Big]^{q/p}. \qquad (2.2.12)$$

Applying Lemma 2.4 to each arc of $\partial \mathbb{D} \setminus E$ separately we get

$$\int_{T_2(h)} \Big(\frac{h f_{\nu+1}(z)}{\rho^{r-\nu}(z)\lambda(\theta;h)} \Big)^p d\theta \leq \int_{\partial \mathbb{D}} \Big(\frac{\Delta^{*}_{f^{(r)}}(e^{i\theta}; 2(r+1)h}{\lambda(\theta;h)} \Big)^p d\theta, \qquad (2.2.13)$$

and (2.2.12) and (2.2.13) yield

$$\int_0^1 \frac{dh}{h^{1+rq}} \Big[\int_{T_2(h)} \Big| \frac{\Sigma_1}{\lambda(\theta;h)} \Big|^p d\theta \Big]^{q/p} < \infty. \qquad (2.2.14)$$

Similarly

$$\int_0^1 \frac{dh}{h^{1+rq}} \Big[\int_{T_2(h)} \Big| \frac{\Sigma_j}{\lambda(\theta;h)} \Big|^p d\theta \Big]^{q/p} < \infty, \quad j=2,3,$$

i.e.

$$\int_0^1 \frac{dh}{h^{1+rq}}\Big[\int_{T_2(h)} |\cdot|^p \, d\theta\Big]^{q/p} < \infty . \tag{2.2.15}$$

$(2.2.2_3)$ We apply as in the case $(2.2.2_2)$ Lemmas 2.4 and 2.5. If $z = e^{i\theta}$, $\zeta = e^{i(\theta+\tau)}$, $0 \leqslant \tau \leqslant (r+1)h$, $\nu \leqslant r-1$, $\zeta^\circ = \left(1 - \frac{1}{\mu_f(z)}\right)\zeta$, then

$$f^{(\nu)}(\zeta) = \sum_{\mu=0}^{r-\nu} \frac{f^{(\nu+\mu)}(\zeta^\circ)}{\mu!}(\zeta-\zeta_0)^\mu + \frac{1}{(r-\nu-1)!}\int_{\zeta^\circ}^{\zeta} (\zeta-t)^{r-\nu-1}(f^{(r)}(t) - f^{(r)}(\zeta^\circ))\,dt,$$

and we have the estimates

$$|f^{(\nu)}(\zeta)| \lesssim \rho(z)^{2(r+1)} + \frac{\Delta^*_{f^{(r)}}(z; 2(r+1)h)}{\mu_f^{r-\nu}(z)}, \quad 0 \leqslant \nu \leqslant r. \tag{2.2.15'}$$

But if $\zeta_1 = e^{i(\theta+\tau)}$, $\zeta_2 = e^{i(\theta+\tau+h)}$, $0 \leqslant \tau \leqslant rh$, $\nu \leqslant r-1$,

then

$$|f^{(\nu)}(\zeta_1) - f^{(\nu)}(\zeta_2)| = \Big| \sum_{\mu=0}^{r-\nu} \frac{f^{(\nu+\mu)}(\zeta_1^\circ)}{\mu!}(\zeta_1 - \zeta_1^\circ)^\mu -$$

$$- \sum_{\mu=0}^{r-\nu} \frac{f^{(\nu+\mu)}(\zeta_2^\circ)}{\mu!}(\zeta_2 - \zeta_2^\circ)^\mu + O\left(\frac{\Delta^*_{f^{(r)}}(z; 2(r+1)h)}{\mu_f(z)^{r-\nu}}\right)\Big| \lesssim$$

$$\lesssim \frac{\Delta^*_{f^{(r)}}(z; 2(r+1)h)}{\mu_f(z)^{r-\nu}} + h\rho^{2(r+1)}(z). \tag{2.2.16}$$

Hence putting $\Delta^* = \Delta^*_{f^{(r)}}(z, H)$, $H = 2(r+1)h$, $\rho = \rho(z)$, $\mu_f = \mu_f(z)$, $a = a(z)$, we get

$$|(f^{(\nu)}(\zeta_1)-f^{(\nu)}(\zeta_2))(d^{\varepsilon})^{(r-\nu)}(\zeta_2)|\leqslant\left(\frac{\Delta^*}{\mu_f^{r-\nu}}+h\rho^{2(r+1)}\right)(\varepsilon^{r-\nu}a^{r-\nu}\rho^{\varepsilon}+\rho^{\frac{\varepsilon}{2}-(r-\nu)}. \qquad (2.2.16')$$

Further

$$\frac{a}{\mu_f}=\log\frac{4}{\rho}\cdot\frac{a}{a_f}\leqslant\log\frac{4}{\rho},\quad \frac{1}{\mu_f}\leqslant\rho,$$

$$(2.2.16')\leqslant\Delta^*+h, \qquad (2.2.17)$$

$$|f^{(\nu)}(\zeta_1)((d^{\varepsilon})^{(r-\nu)}(\zeta_2)-(d^{\varepsilon})^{(r-\nu)}(\zeta_1))|\leqslant$$

$$\leqslant(\rho^{2(r+1)}+\frac{\Delta^*}{\mu_f^{r-\nu}})|(d^{\varepsilon})^{(r-\nu)}(\zeta_1)-(d^{\varepsilon})^{(r-\nu)}(\zeta_2)|\leqslant$$

$$\leqslant(\rho^{2(r+1)}+\frac{\Delta^*}{\mu_f^{r-\nu}})\cdot h\cdot(\varepsilon^{r-\nu+1}a^{r-\nu+1}\rho^{\varepsilon}+\rho^{\frac{\varepsilon}{2}-(r-\nu+1)})\leqslant h+\Delta^*, \qquad (2.2.18)$$

$$h|(fd^{\varepsilon})^{(r)}(\zeta_1)|\leqslant h+\Delta^*. \qquad (2.2.18')$$

Taking (2.2.16') - (2.2.18') into account we obtain

$$\int_0^1\frac{dh}{h^{1+rq}}\left[\int_{\arg T_3(h)}|\cdot|^p d\theta\right]^{q/p}<\infty. \qquad (2.2.19)$$

($2.2.2_4$) We need analogous of the estimate (2.2.15'). Let us introduce the convex region $G_H(z)=\bigcup_{0\leqslant r\leqslant 1}D(rz,H)$;
due to subadditivity of the moduli of continuity for arbitrary points ζ, $\zeta^0\in G_H(z)$ and for any continuous function

we have the following relation

$$|\varphi(\zeta)-\varphi(\zeta^\circ)| \leqslant 10\left(1+\frac{|\zeta-\zeta_0|}{H}\right)\Delta^*_\varphi(z;H). \tag{2.2.20}$$

Consequently we can write down for our function $\left(\zeta^\circ=\left(1-\frac{1}{\mu_f(z)}\right)\zeta\right)$

$$|f^{(\nu)}(\zeta)|=\left|\sum_{\mu=0}^{r-\nu}\frac{f^{(\nu+\mu)}(\zeta^\circ)}{\mu!}(\zeta-\zeta^\circ)^{\mu}+\frac{1}{(r-\nu-1)!}\int_{\zeta_0}^{\zeta}(\zeta-t)^{r-\nu-1}\left(f^{(r)}(t)-f^{(r)}(\zeta_0^\circ)\right)dt\right| \lesssim$$

$$\lesssim \rho^{2(r+1)}(z)+\Delta^*\cdot\int_{\zeta^\circ}^{\zeta}|\zeta-t|^{r-\nu-1}\left(1-\frac{|t-\zeta^\circ|}{h}\right)|dt| \lesssim$$

$$\lesssim \rho^{2(r+1)}+\left(\frac{1}{\mu_f^{r-\nu}}+\frac{1}{\mu_f^{r-\nu+1}h}\right)\Delta^* \lesssim \rho^{2(r+1)}+\frac{\Delta^*}{\mu_f^{r-\nu+1}h}, \tag{2.2.21}$$

$$\Delta^*=\Delta^*_{f^{(r)}}(z;H), \quad H=2(r+1)h, \quad \rho=\rho(z), \quad \mu_f=\mu_f(z),$$

and

$$|f^{(r)}(\zeta)| \leqslant |f^{(r)}(\zeta^\circ)|+|f^{(r)}(\zeta)-f^{(r)}(\zeta^\circ)| \lesssim \rho^{2(r+1)}+\frac{\Delta^*}{\mu_f h}. \tag{2.2.22}$$

Keeping the notations of $(2.2.2_3)$ we have by Lemma 2.4

$$|(d^2)^{(\nu)}-(d^\varepsilon)^{(\nu)}(\zeta_2)| \lesssim h\left(\varepsilon^{\nu+1}a(z)^{\nu+1}\rho(z)^{\varepsilon}+\rho(z)^{\frac{\varepsilon}{2}-\nu-1}\right). \tag{2.2.23}$$

Now with the help of (2.2.20) and Lemma 2.5 we get

$$|f^{(r)}(\zeta_1)-f^{(r)}(\zeta_2)| \le \Delta^* = \Delta^*_{f^{(r)}}(z;2(r+1)h),$$

$$|f^{(\nu)}(\zeta_1)-f^{(\nu)}(\zeta_2)| \le \Big|\sum_{\mu=0}^{r-\nu}\Big[\frac{f^{(\nu+\mu)}(\zeta_1^0)}{\mu!}(\zeta_1-\zeta_1^0)^{\mu}-\frac{f^{(\nu+\mu)}(\zeta_1^0)}{\mu!}(\zeta_2-\zeta_1^0)^{\mu}\Big]\Big| +$$

$$+\frac{1}{(r-\nu-1)!}\Big|\int_{\zeta_1^0}^{\zeta_1}(\zeta_1-t)^{r-\nu-1}(f^{(r)}(t)-f^{(r)}(\zeta_1^0))\,dt -$$

$$-\int_{\zeta_1^0}^{\zeta_2}(\zeta_2-\tau)^{r-\nu-1}(f^{(r)}(\tau)-f^{(r)}(\zeta_1^0))\,d\tau\Big| \le h\rho^{2(r+1)} +$$

$$+\Big|\int_{\zeta_1^0}^{\zeta_1}(\zeta_1-t)^{r-\nu-1}(f^{(r)}(t)-f^{(r)}(\zeta_1^0))\,dt-\int_{\zeta_1^0}^{\zeta_2}(\zeta_1-t)^{r-\nu-1}(f^{(r)}(t)-f^{(r)}(\zeta_1^0))\,dt -$$

$$-\int_{\zeta_1^0}^{\zeta_2}[(\zeta_2-t)^{r-\nu-1}-(\zeta_1-t)^{r-\nu-1}](f^{(r)}(t)-f^{(r)}(\zeta_1^0))\,dt\Big| \le$$

$$\le h\rho^{2(r+1)}+h\cdot\frac{\Delta^*}{\mu_f^{r-\nu}h}+\Big|\int_{\zeta_1}^{\zeta_2}(\zeta_1-t)^{r-\nu-1}(f^{(r)}(t)-f^{(r)}(\zeta_1^0))\,dt\Big| \le$$

$$\le h\rho^{2(r+1)}+\frac{\Delta^*}{\mu_f^{r-\nu}}+\frac{h^{r-\nu-1}\Delta^*}{\mu_f} \le h\rho^{2(r+1)}+\frac{\Delta^*}{\mu_f^{r-\nu}}, \quad 0\le\nu\le r-1. \tag{2.2.24}$$

Hence (2.2.21) - (2.2.24) yield

$$|(f^{(\nu)}(\zeta_1)-f^{(\nu)}(\zeta_2))(d^{\varepsilon})^{(r-\nu)}(\zeta_1)| \le \Big(\frac{\Delta^*}{\mu_f^{r-\nu}}+h\rho^{2(r+1)}\Big)(\varepsilon^{r-\nu}a^{r-\nu}\rho^{\varepsilon}+\rho^{\frac{\varepsilon}{2}-r+\nu}) \le$$

$$\le \Delta^*+h, \tag{2.2.25}$$

$$|f^{(\nu)}(\zeta_2)((d^\varepsilon)^{(r-\nu)}(\zeta_2)-(d^\varepsilon)^{(r-\nu)}(\zeta_1))| \leqslant (\rho^{2(r+1)}-\frac{\Delta^*}{\mu_f^{r-\nu+1}h})\cdot h\cdot(\varepsilon^{r-\nu+1}a^{r-\nu+1}\rho^\varepsilon +$$

$$+\rho^{\frac{\varepsilon}{2}-r+\nu-1}) \leqslant \Delta^* + h, \tag{2.2.26}$$

$$h\cdot|(fd^\varepsilon)^{(r)}(\zeta_1)| \leqslant h+\Delta^*. \tag{2.2.27}$$

Finally (2.2.25) - (2.2.27) imply

$$\int_0^1 \frac{dh}{h^{1+rq}}\Big[\int_{\arg T_4(h)} |\cdot|^p\, d\theta\Big]^{q/p} < \infty. \tag{2.2.28}$$

Gathering the estimates (2.2.4), (2.2.15), (2.2.19), (2.2.28) together we finish the proof of Statement 1.

2.3. PROOF OF STATEMENT 2.

In this situation we divide the circle $\partial\mathbb{D}$ onto following seven sets:

$$T_1(h) = \{z\in\partial\mathbb{D} : \rho(z)\leqslant 2(r+1)h\};$$

$$T_2(h) = \{z\in\partial\mathbb{D}\setminus\mathfrak{T} : \rho(z)>2(r+1)h,\quad a_f(z)\leqslant A\,\frac{\log\frac{4}{\rho(z)}}{\rho(z)}\};$$

$$T_3(h) = \{z\in\partial\mathbb{D}\setminus\mathfrak{T} : \rho(z)>2(r+1)h,\quad a_f(z)> A\,\frac{\log\frac{4}{\rho(z)}}{\rho(z)},\ h\geqslant\frac{1}{\mu_f(z)}\};$$

$$T_4(h) = \{z\in\partial\mathbb{D}\setminus\mathfrak{T} : \rho(z)>2(r+1)h,\quad a_f(z)> A\,\frac{\log\frac{4}{\rho(z)}}{\rho(z)},\ h<\frac{1}{\mu_f(z)}\};$$

$$T_{2,\tau}(h)=\{z\in\tau:\rho(z)>2(r+1)h,\ a_f(z)\leqslant A\frac{\log\frac{4}{\rho(z)}}{\rho(z)}\};$$

$$T_{3,\tau}(h)=\{z\in\tau:\rho(z)>2(r+1)h,\ a_f(z)>A\frac{\log\frac{4}{\rho(z)}}{\rho(z)},\ h\geqslant\frac{1}{\mu_f(z)}\};$$

$$T_{4,\tau}(h)=\{z\in\tau:\rho(z)>2(r+1)h,\ a_f(z)>A\frac{\log\frac{4}{\rho(z)}}{\rho(z)},\ h<\frac{1}{\mu_f(z)}\}.$$

The estimates of terms $T_1(h) - T_4(h)$ are similar to the estimates in $(2.2.2_1)$ - $(2.2.2_4)$ of Statement 1 with the use of the relation

$$|(F_\tau^N)^{(\nu)}(\zeta)|\leqslant a_f^\nu(z),\quad \nu\geqslant 1,\quad z=e^{i\theta}\in\partial\mathbb{D}\setminus(\tau\cup T_1(h)),\ \zeta=e^{i(\theta+\tau)},$$
$$0\leqslant\tau\leqslant rh,$$

instead of the difference $(fd^\varepsilon)^{(r)}(\zeta_1)-(fd^\varepsilon)^{(r)}(\zeta_2)$ it is to consider the difference $(fd^\varepsilon F_\tau^N)^r(\zeta_1)-(fd^\varepsilon F_\tau^N)^{(r)}(\zeta_2)$ for

$$\zeta_1=e^{i(\theta+\tau)},\quad \zeta_2=e^{i(\theta+\tau+h)},\quad 0\leqslant\tau\leqslant rh,$$

$$e^{i\theta}=z\in\partial\mathbb{D}\setminus(\tau\cup T_1(h)),$$

and that reduces to the estimate of summands of the following three types:

$$(f^{(\nu)}(\zeta_1)-f^{(\nu)}(\zeta_2))(d^\varepsilon)^{(\mu)}(\zeta_1)(F_\tau^N)^{(r-\mu-\nu)}(\zeta_1),\quad 0\leqslant\mu+\nu\leqslant r,$$

$$f^{(\nu)}(\zeta_2)((d^\varepsilon)^{(\mu)}(\zeta_1)-(d^\varepsilon)^{(\mu)}(\zeta_2))(F_\tau^N)^{(r-\mu-\nu)}(\zeta_1),$$

$$f^{(\nu)}(\zeta_2)(d^{\varepsilon})^{(\mu)}(\zeta_2)((F_{\tau}^{N})^{(r-\mu-\nu)}(\zeta_1)-(F_{\tau}^{N})^{(r-\mu-\nu)}(\zeta_2)).$$

That estimates it is easy to accomplish literally as in the proof of Statement 1. Some modifications are required only by the consideration of the case $z \in \tau \setminus T_1(h)$. Let ψ be the outer part of f_0. Lemma 2.6 implies $\psi \in X^{r}_{pq}(\omega, l)$. Let us consider the following outer function

$$|\Phi_{\tau}(z)| = \begin{cases} 1, & z \in \tau, \\ |f_0(z)|, & z \in \partial\mathbb{D} \setminus \tau. \end{cases}$$

Then $\psi = F_{\tau}\Phi_{\tau}$ and further

$$f F_{\tau}^{N} d^{\varepsilon} = (f\psi^{N})\,\Phi_{\tau}^{-N} d^{\varepsilon}.$$

For $z \in \tau$ we can write the estimates

$$|(\Phi_{\tau}^{-N})^{(\nu)}(z)| \preccurlyeq a_f^{\nu}(z), \quad \nu \geqslant 1,$$

i.e. it is possible to study the cases $T_{2,\tau}(h) - T_{4,\tau}(h)$ like $T_2(h) - T_4(h)$ taking the function f instead of $f\psi^{N} \in X$. That completes the proof of Statement 2 and hence of Theorem 14. ●

REFERENCES

1. Shilov G.E. Rings of functions with the uniform convergence, Ukr. Mat.Zh.,1951, 4, N 4, 404-411.
2. Carleson L. Sets of uniqueness for function regular in the unit circle.- Acta Math., 1952, 87, N 3-4, 325-345.
3. Carleson L. A representation formula for the Dirichlet integral.- Math.Z., 1960, 73, N 2, 190-196.
4. Havin V.P. A factorization of analytic functions which are smooth up to the boundary.- Zap.Nauchn.Semin. Lening.Otd.Mat.Inst. Steklova, 1971, 22, 202-205.
5. Rudin W. The closed ideals in an algebra of analytic functions.- Canad.J.Math., 1957, 9, N 3, 426-434.
6. Korenblum B.I., Korolevich V.S. Analytical functions which are regular in the disc and smooth up to the boundary.- Mat. Zametki , 1970, 7, N 2, 165-172.
7. Korenblum B.I., An extremal property for outer functions.- Mat. Zametki, 1971, 10, N 1, 53-56.
8. Shamoyan F.A. The division by inner function in some spaces of analytic functions.- Zap.Nauchn.Semin. Lening.Otd.Mat.Inst. Steklova, 1971, 22, 206-208.
9. Gurarii V.P. A factorization of absolutely convergent Taylor series and Fourier-integrals.- Zap.Nauchn.Semin.Lening.Otd.Mat.Inst.Steklova, 1972, 30, 15-32.
10. Verbitskii I.E. Multiplicators in ℓ^p_A -spaces.- Funk.Anal. i Prilozh., 1980. 14, N 3, 67-68.
11. Anderson J.M. Algebras contained within H^∞ .- Zap.Nauchn.Semin. Lening.Otd.Mat.Inst.Steklova, 1978, 81, 235-236.
12. Havin V.P., Shamoyan F.A. Analytic functions with the Lipschitz module of their boundary values.- Zap.Nauchn.Semin.Lening.Otd.Mat. Inst.Steklova, 1970, 19, 237-239.
13. Havin V.P. A generalization of the Zygmund-Privalov theorem about modulus of continuity of the conjugate function.-Izv.Akad.Nauk

Arm.SSR, Mat., 1971, 6, N 2-3, 252-258; N 4, 265-287.

14. Brennan J. Approximation in the mean by polynomials on non Carathéodory domains.- Ark.Mat., 1977, 15, N 1, 117-168.

15. Taylor B.A., Williams D.L. Zeros of Lipschitz functions analytic in the unit disc.- Mich.Math.J., 1971, 18, N 2, 129-139.

16. Korenblum B.I. Functions holomorphic in the disc and smooth up to its boundary.- Dokl.Akad.Nauk SSSR, 1971, 200, N 1, 24-27.

17. Stegbuchner J. Nullstellen analytischen Functionen und verallgemeinerte Carleson-Mengen, I.- Sitznugsber. Öster, Akad. Wiss., Math.-Nat., 1975, Abt.2, 183, N 8-10, 463-503.

18. Stegbuchner J. ε-Invarianten bei verallgemeinerten Carleson-Mengen.- Monats hefte.Math., 1976, 81, N 3, 217-224.

19. Schwartz L. Etudes des sommes d'exponentielles reelles. Paris, 1943.

20. Hirschmen I.I., Jenkus J.A. On lacunary Dirichlet series.- Proc. Amer.Math.Soc., 1950, 1, N 4, 512-517.

21. Anderson J.M. Bounded analytic functions with Hadamard gaps.- Mathematica, 1976, 23, N 2, 142-147.

22. Gurarii V.P. The spectral synthesis of functions bounded on a half axis.- Funk.Anal. i Prilozh., 1969, 3, N 4, 34-46.

23. Korenblum B.I. The closed ideals of the ring A^n .- Funk.Anal. i Prilozh, 1972, 6, N 3, 38-53.

24. Korenblum B.I. Invariant subspaces of the shift-operator on a weihted Hilbert-space.- Mat-Sb., Nov.Ser., 1972, 89, N 1, 110-137.

25. Shamoyan F.A. The constructing of a special sequence and the closed ideals-structure in some algebras of analytic functions.- Izv.Akad.Nauk Arm.SSR,Mat., 1972, 7, N 6, 440-470.

26. Shamoyan F.A. The closed ideals-structure in some algebras of analytic functions smooth up to the boundary.- Dokl.Akad.Nauk Arm. SSR, 1975, 60, N 3, 133-136.

27. Shamoyan F.A. The closed ideals in algebras of analytic functions smooth up to the boundary.- Izv.Akad.Arm.SSR, Mat., 1981, 16, N 3, 173-191.

28. Tamrazov P.M. Contour and solid structure properties of holomorphic functions of a complex variable.- Usp.Mat.Nauk, 1973, 28, N 1, 131-161.

29. Dyn'kin E.M. Estimates of analytic functions in Jordan regions.- Zap.Nauchn.Semin.Lening.Otd.Mat.Inst.Steklova, 1977, 73, 70-90.

30. MucKenhoupt B. Weighted norm inequalities for the Hardy maximal functions.- Trans.Amer.Math.Soc., 1972, 165, 207-226.

31. Zygmund A. Trigonometric series, v.I, Moskva, "Mir", 1965.

32. Dziadyk V.K. Introduction in the theory of the uniform approximation by polynomials, Moskva, 2Nauka", 1977.

33. Goluzin G.M. The geometric theory of functions of a complex variable, Moskva, "Nauka", 1966.

34. Stein E.M. Singular integrals and differentiality properties of functions, Princeton, 1970.

35. Gol'dberg A.A., Ostrovskii I.V. The distribution of values of meromorphic functions, Moskva, "Nauka", 1970.

36. Levin B.Ya. The distribution of zeros of entire functions, Moskva, 1956.

37. Dyn'kin E.M. The constructive characteristic of S.L.Sobolev and O.V.Besov classes.- Tr.Mat.Inst.Steklova, 1981, 155, 41-76.

38. Krein S.G., Petunin Yu.I., Semenov E.M. Interpolation of linear operators. Moskva, "Nauka", 1978.

39. Dyn'kin E.M. Smooth functions on plain sets.- Dokl.Akad. Nauk SSSR, 1979, 208, N 1, 25-27.

40. Dzhrbashian M.M. Integral transforms and representations of functions in a complex domain, Moskva, "Nauka", 1966.

41. Shamoyan F.A. A class of Toeplitz-operators which is connected with the division-property of analytic functions.- Funk.Anal. i Prilozh., 1979, 13, N 1, 83-84.

42. Rubel L.A. A generalized canonical product "Contemporary problems of analytic functions", Moskva, "Nauka", 1965, 264-269.

43. Horowitz C. Factorization theorems for functions in the Bergman spaces.- Duke Math.J., 1977, 44, N 1, 201-213.

44. Privalov I.I. The boundary properties of analytic functions, Moskva, 1950.

45. Hoffman K. Banach spaces of analytic functions. Prentice-Hall, Eugl.diffs, N.Z., 1962.

46. Kahanu J.-P. Best approximation in $L^1(\mathbb{T})$.- Bull.Amer.Math.

Soc., 1974, 80, N 5, 788-804.

47. Bruna J., Ortega J. Closed finitely generated ideals in algebras of holomorphic functions and smooth to the boundary in strictly pseudoconvex domains.- Math.Ann., 1984, 268, N 2, 137-157.

48. Shamoyan F.A. Toeplitz-operators and the division by inner function in some spaces of analytic functions.- Dokl.Akad.Nauk Arm. SSR, 1983, 76, N 3, 109-113.

49. Vinogradov S.A., Shirokov N.A. A factorization of analytic functions with the derivative from H^p .- Zap.Nauchn.Semin.Lening. Otd.Mat.Inst.Steklova, 1971, 22, 8-27.

50. Shirokov N.A. Properties of primary ideals of absolutely convergent Taylor-series and Fourier-integrals.- Zap.Nauchn.Semin.Lening. Otd.Mat.Inst.Steklova, 1974, 32, 149-161.

51. Shirokov N.A. Ideals and factorization in algebras of analytic functions smooth up to the boundary.- Tr.Mat.Inst.Steklova, 1978, 130, 196-222.

52. Shirokov N.A. Standard ideals of the algebra H^1_n .- Funk.Anal. i Prilozh., 1979, 13, N 1, 86-87.

53. Shirokov N.A. Division and multiplication by inner function in spaces of analytic functions smooth up to the boundary.- Lect. Notes in Math., 1981, 864, 413-440.

54. Shirokov N.A. A module of boundary values of analytic functions from Λ^n_ω .- Zap.Nauchn.Semin.Lening.Otd.Mat.Inst.Steklova, 1981, 113, 258-260.

55. Shirokov N.A. Zero sets of a function from Λ_ω .- Zap.Nauchn. Semin.Lening.Otd.Mat.Inst.Steklova, 1982, 107, 178-188.

56. Shirokov N.A. The division by inner function does not reduce the smoothness rate.- Dokl.Akad. Nauk SSSR, 1983, 268, N 4, 821-823.

57. Shirokov N.A. Some properties of series with small gaps.- Usp.Mat. Nauk, 1982, 37, N 2, 249-250.

58. Shirokov N.A. Moduli of analytic functions smooth up to the boundary, LOMI-Preprint, 1982, R-7-82.

59. Shirokov N.A. Closed ideals of algebras of B^α_{pq} -type. Izv.Akad.Nauk SSSR,Mat., 1982, 46, N 6, 1316-1333.

60. Shirokov N.A. The Nevanlinna-factorization in some classes of analytic functions.-Zap.Nauchn.Semin.Lening.Otd.Mat.Inst.Steklova,

1983, 126, 205-207.

61. Shirokov N.A. Properties of the module of analytic functions smooth up to the boundary.- Dokl.Akad.Nauk SSSR, 1983, 269, N 6, 1320-1323.

SUBJECT INDEX

LECTURE NOTES IN MATHEMATICS

Edited by A. Dold and B. Eckmann

Some general remarks on the publication of monographs and seminars

In what follows all references to monographs, are applicable also to multiauthorship volumes such as seminar notes.

Lecture Notes aim to report new developments - quickly, informally, and at a high level. Monograph manuscripts should be reasonably self-contained and rounded off. Thus they may, and often will, present not only results of the author but also related work by other people. Furthermore, the manuscripts should provide sufficient motivation, examples and applications. This clearly distinguishes Lecture Notes manuscripts from journal articles which normally are very concise. Articles intended for a journal but too long to be accepted by most journals, usually do not have this "lecture notes" character. For similar reasons it is unusual for Ph.D. theses to be accepted for the Lecture Notes series.

Experience has shown that English language manuscripts achieve a much wider distribution.

Manuscripts or plans for Lecture Notes volumes should be submitted either to one of the series editors or to Springer-Verlag, Heidelberg. These proposals are then refereed. A final decision concerning publication can only be made on the basis of the complete manuscripts, but a preliminary decision can usually be based on partial information: a fairly detailed outline describing the planned contents of each chapter, and an indication of the estimated length, a bibliography, and one or two sample chapters - or a first draft of the manuscript. The editors will try to make the preliminary decision as definite as they can on the basis of the available information.

Lecture Notes are printed by photo-offset from typed copy delivered in camera-ready form by the authors. Springer-Verlag provides technical instructions for the preparation of manuscripts, and will also, on request, supply special stationery on which the prescribed typing area is outlined. Careful preparation of manuscripts will help keep production time short and ensure satisfactory appearance of the finished book. Running titles are not required; if however they are considered necessary, they should be uniform in appearance. We generally advise authors not to start having their final manuscripts specially typed beforehand. For professionally typed manuscripts, prepared on the special stationery according to our instructions, Springer-Verlag will, if necessary, contribute towards the typing costs at a fixed rate.

The actual production of a Lecture Notes volume takes 6-8 weeks.

LECTURE NOTES IN MATHEMATICS
Edited by A. Dold and B. Eckmann

Some general remarks on the publication of monographs and seminars

In what follows all references to monographs, are applicable also to multiauthorship volumes such as seminar notes.

1. Lecture Notes aim to report new developments - quickly, informally, and at a high level. Monograph manuscripts should be reasonably self-contained and rounded off. Thus they may, and often will, present not only results of the author but also related work by other people. Furthermore, the manuscripts should provide sufficient motivation, examples and applications. This clearly distinguishes Lecture Notes manuscripts from journal articles which normally are very concise. Articles intended for a journal but too long to be accepted by most journals, usually do not have this "lecture notes" character. For similar reasons it is unusual for Ph.D. theses to be accepted for the Lecture Notes series.

   Experience has shown that English language manuscripts achieve a much wider distribution.

2. Manuscripts or plans for Lecture Notes volumes should be submitted either to one of the series editors or to Springer-Verlag, Heidelberg. These proposals are then refereed. A final decision concerning publication can only be made on the basis of the complete manuscripts, but a preliminary decision can usually be based on partial information: a fairly detailed outline describing the planned contents of each chapter, and an indication of the estimated length, a bibliography, and one or two sample chapters - or a first draft of the manuscript. The editors will try to make the preliminary decision as definite as they can on the basis of the available information.

3. Lecture Notes are printed by photo-offset from typed copy delivered in camera-ready form by the authors. Springer-Verlag provides technical instructions for the preparation of manuscripts, and will also, on request, supply special staionery on which the prescribed typing area is outlined. Careful preparation of the manuscripts will help keep production time short and ensure satisfactory appearance of the finished book. Running titles are not required; if however they are considered necessary, they should be uniform in appearance. We generally advise authors not to start having their final manuscripts specially tpyed beforehand. For professionally typed manuscripts, prepared on the special stationery according to our instructions, Springer-Verlag will, if necessary, contribute towards the typing costs at a fixed rate.

   The actual production of a Lecture Notes volume takes 6-8 weeks.

.../...

4. Final manuscripts should contain at least 100 pages of mathematical text and should include

   - a table of contents
   - an informative introduction, perhaps with some historical remarks. It should be accessible to a reader not particularly familiar with the topic treated.
   - subject index; this is almost always genuinely helpful for the reader.

5. Authors receive a total of 50 free copies of their volume, but no royalties. They are entitled to purchase further copies of their book for their personal use at a discount of 33 1/3 %, other Springer mathematics books at a discount of 20 % directly from Springer-Verlag.

   Commitment to publish is made by letter of intent rather than by signing a formal contract. Springer-Verlag secures the copyright for each volume.

LECTURE NOTES

ESSENTIALS FOR THE PREPARATION OF CAMERA-READY MANUSCRIPTS

The preparation of manuscripts which are to be reproduced by photo-offset requires special care. Manuscripts which are submitted in technically unsuitable form will be returned to the author for retyping. There is normally no possibility of carrying out further corrections after a manuscript is given to production. Hence it is crucial that the following instructions be adhered to closely. If in doubt, please send us 1 - 2 sample pages for examination.

Typing area. On request, Springer-Verlag will supply special paper with the typing area outlined.

The CORRECT TYPING AREA is 18 x 26 1/2 cm (7,5 x 11 inches).

Make sure the TYPING AREA IS COMPLETELY FILLED. Set the margins so that they precisely match the outline and type right from the top to the bottom line. (Note that the page-number will lie outside this area). Lines of text should not end more than three spaces inside or outside the right margin (see example on page 4).

Type on one side of the paper only.

Type. Use an electric typewriter if at all possible. CLEAN THE TYPE before use and always use a BLACK ribbon (a carbon ribbon is best).

Choose a type size large enough to stand reduction to 75%.

Word Processors. Authors using word-processing or computer-typesetting facilities should follow these instructions with obvious modifications. Please note with respect to your printout that
i) the characters should be sharp and sufficiently black;
ii) if the size of your characters is significantly larger or smaller than normal typescript characters, you should adapt the length and breadth of the text area proportionally keeping the proportions 1:0.68.
iii) it is not necessary to use Springer's special typing paper. Any white paper of reasonable quality is acceptable.
IF IN DOUBT, PLEASE SEND US 1-2 SAMPLE PAGES FOR EXAMINATION. We will be glad to give advice.

Spacing and Headings (Monographs). Use ONE-AND-A-HALF line spacing in the text. Please leave sufficient space for the title to stand out clearly and do NOT use a new page for the beginning of subdivisions of chapters. Leave THREE LINES blank above and TWO below headings of such subdivisions.

Spacing and Headings (Proceedings). Use ONE-AND-A-HALF line spacing in the text. Start each paper on a NEW PAGE and leave sufficient space for the title to stand out clearly. However, do NOT use a new page for the beginning of subdivisions of a paper. Leave THREE LINES blank above and TWO below headings of such subdivisions. Make sure headings of equal importance are in the same form.

The first page of each contribution should be prepared in the same way. Therefore, we recommend that the editor prepares a sample page and passes it on to the authors together with these ESSENTIALS. Please take

.../...

the following as an example.

MATHEMATICAL STRUCTURE IN QUANTUM FIELD THEORY

John E. Robert
Fachbereich Physik, Universität Osnabrück
Postfach 44 69, D-4500 Osnabrück

Please leave THREE LINES blank below heading and address of the author. THEN START THE ACTUAL TEXT OF YOUR CONTRIBUTION.

Footnotes: These should be avoided. If they cannot be avoided, place them at the foot of the page, separated from the text by a line 4 cm long, and type them in SINGLE LINE SPACING to finish exactly on the outline.

Symbols. Anything which cannot be typed may be entered by hand in BLACK AND ONLY BLACK ink. (A fine-tipped rapidograph is suitable for this purpose; a good black ball-point will do, but a pencil will not). Do not draw straight lines by hand without a ruler (not even in fractions).

Equations and Computer Programs. Equations and computer programs should begin four spaces inside the left margin. Should the equations be numbered, then each number should be in brackets at the right-hand edge of the typing area.

Pagination. Number pages in the upper right-hand corner in LIGHT BLUE OR GREEN PENCIL ONLY. The final page numbers will be inserted by the printer.

There should normally be NO BLANK PAGES in the manuscript (between chapters or between contributions) unless the book is divided into Part A, Part B for example, which should then begin on a right-hand page.

It is much safer to number pages AFTER the text has been typed and corrected. Page 1 (Arabic) should be THE FIRST PAGE OF THE ACTUAL TEXT. The Roman pagination (table of contents, preface, abstract, acknowledgements, brief introductions, etc.) will be done by Springer-Verlag.

Corrections. When corrections have to be made, cut the new text to fit and PASTE it over the old. White correction fluid may also be used.

Never make corrections or insertions in the text by hand.

If the typescript has to be marked for any reason, e.g. for TEMPORARY page numbers or to mark corrections for the typist, this can be done VERY FAINTLY with BLUE or GREEN PENCIL but NO OTHER COLOR: these colors do not appear after reproduction.

Table of Contents. It is advisable to type the table of contents later, copying the titles from the text and inserting page numbers.

Literature References. These should be placed at the end of each paper or chapter, or at the end of the work, as desired. Type them with single line spacing and start each reference on a new line.
Please ensure that all references are COMPLETE and PRECISE.

Vol. 1145: G. Winkler, Choquet Order and Simplices. VI, 143 pages. 1985.

Vol. 1146: Séminaire d'Algèbre Paul Dubreil et Marie-Paule Malliavin. Proceedings, 1983–1984. Edité par M.-P. Malliavin. IV, 420 pages. 1985.

Vol. 1147: M. Wschebor, Surfaces Aléatoires. VII, 111 pages. 1985.

Vol. 1148: Mark A. Kon, Probability Distributions in Quantum Statistical Mechanics. V, 121 pages. 1985.

Vol. 1149: Universal Algebra and Lattice Theory. Proceedings, 1984. Edited by S. D. Comer. VI, 282 pages. 1985.

Vol. 1150: B. Kawohl, Rearrangements and Convexity of Level Sets in PDE. V, 136 pages. 1985.

Vol 1151: Ordinary and Partial Differential Equations. Proceedings, 1984. Edited by B.D. Sleeman and R.J. Jarvis. XIV, 357 pages. 1985.

Vol. 1152: H. Widom, Asymptotic Expansions for Pseudodifferential Operators on Bounded Domains. V, 150 pages. 1985.

Vol. 1153: Probability in Banach Spaces V. Proceedings, 1984. Edited by A. Beck, R. Dudley, M. Hahn, J. Kuelbs and M. Marcus. VI, 457 pages. 1985.

Vol. 1154: D.S. Naidu, A.K. Rao, Singular Pertubation Analysis of Discrete Control Systems. IX, 195 pages. 1985.

Vol. 1155: Stability Problems for Stochastic Models. Proceedings, 1984. Edited by V.V. Kalashnikov and V.M. Zolotarev. VI, 447 pages. 1985.

Vol. 1156: Global Differential Geometry and Global Analysis 1984. Proceedings, 1984. Edited by D. Ferus, R.B. Gardner, S. Helgason and U. Simon. V, 339 pages. 1985.

Vol. 1157: H. Levine, Classifying Immersions into $\mathbb{R}^4$ over Stable Maps of 3-Manifolds into $\mathbb{R}^2$. V, 163 pages. 1985.

Vol. 1158: Stochastic Processes – Mathematics and Physics. Proceedings, 1984. Edited by S. Albeverio, Ph. Blanchard and L. Streit. VI, 230 pages. 1986.

Vol. 1159: Schrödinger Operators, Como 1984. Seminar. Edited by S. Graffi. VIII, 272 pages. 1986.

Vol. 1160: J.-C. van der Meer, The Hamiltonian Hopf Bifurcation. VI, 115 pages. 1985.

Vol. 1161: Harmonic Mappings and Minimal Immersions, Montecatini 1984. Seminar. Edited by E. Giusti. VII, 285 pages. 1985.

Vol. 1162: S.J.L. van Eijndhoven, J. de Graaf, Trajectory Spaces, Generalized Functions and Unbounded Operators. IV, 272 pages. 1985.

Vol. 1163: Iteration Theory and its Functional Equations. Proceedings, 1984. Edited by R. Liedl, L. Reich and Gy. Targonski. VIII, 231 pages. 1985.

Vol. 1164: M. Meschiari, J.H. Rawnsley, S. Salamon, Geometry Seminar "Luigi Bianchi" II – 1984. Edited by E. Vesentini. VI, 224 pages. 1985.

Vol. 1165: Seminar on Deformations. Proceedings, 1982/84. Edited by J. Ławrynowicz. IX, 331 pages. 1985.

Vol. 1166: Banach Spaces. Proceedings, 1984. Edited by N. Kalton and E. Saab. VI, 199 pages. 1985.

Vol. 1167: Geometry and Topology. Proceedings, 1983–84. Edited by J. Alexander and J. Harer. VI, 292 pages. 1985.

Vol. 1168: S.S. Agaian, Hadamard Matrices and their Applications. III, 227 pages. 1985.

Vol. 1169: W.A. Light, E.W. Cheney, Approximation Theory in Tensor Product Spaces. VII, 157 pages. 1985.

Vol. 1170: B.S. Thomson, Real Functions. VII, 229 pages. 1985.

Vol. 1171: Polynômes Orthogonaux et Applications. Proceedings, 1984. Edité par C. Brezinski, A. Draux, A.P. Magnus, P. Maroni et A. Ronveaux. XXXVII, 584 pages. 1985.

Vol. 1172: Algebraic Topology, Göttingen 1984. Proceedings. Edited by L. Smith. VI, 209 pages. 1985.

Vol. 1173: H. Delfs, M. Knebusch, Locally Semialgebraic Spaces. XVI, 329 pages. 1985.

Vol. 1174: Categories in Continuum Physics, Buffalo 1982. Seminar. Edited by F.W. Lawvere and S.H. Schanuel. V, 126 pages. 1986.

Vol. 1175: K. Mathiak, Valuations of Skew Fields and Projective Hjelmslev Spaces. VII, 116 pages. 1986.

Vol. 1176: R.R. Bruner, J.P. May, J.E. McClure, M. Steinberger, H_∞ Ring Spectra and their Applications. VII, 388 pages. 1986.

Vol. 1177: Representation Theory I. Finite Dimensional Algebras. Proceedings, 1984. Edited by V. Dlab, P. Gabriel and G. Michler. XV, 340 pages. 1986.

Vol. 1178: Representation Theory II. Groups and Orders. Proceedings, 1984. Edited by V. Dlab, P. Gabriel and G. Michler. XV, 370 pages. 1986.

Vol. 1179: Shi J.-Y. The Kazhdan-Lusztig Cells in Certain Affine Weyl Groups. X, 307 pages. 1986.

Vol. 1180: R. Carmona, H. Kesten, J.B. Walsh, École d'Été de Probabilités de Saint-Flour XIV – 1984. Édité par P.L. Hennequin. X, 438 pages. 1986.

Vol. 1181: Buildings and the Geometry of Diagrams, Como 1984. Seminar. Edited by L. Rosati. VII, 277 pages. 1986.

Vol. 1182: S. Shelah, Around Classification Theory of Models. VII, 279 pages. 1986.

Vol. 1183: Algebra, Algebraic Topology and their Interactions. Proceedings, 1983. Edited by J.-E. Roos. XI, 396 pages. 1986.

Vol. 1184: W. Arendt, A. Grabosch, G. Greiner, U. Groh, H.P. Lotz, U. Moustakas, R. Nagel, F. Neubrander, U. Schlotterbeck, One-parameter Semigroups of Positive Operators. Edited by R. Nagel. X, 460 pages. 1986.

Vol. 1185: Group Theory, Beijing 1984. Proceedings. Edited by Tuan H.F. V, 403 pages. 1986.

Vol. 1186: Lyapunov Exponents. Proceedings, 1984. Edited by L. Arnold and V. Wihstutz. VI, 374 pages. 1986.

Vol. 1187: Y. Diers, Categories of Boolean Sheaves of Simple Algebras. VI, 168 pages. 1986.

Vol. 1188: Fonctions de Plusieurs Variables Complexes V. Séminaire, 1979–85. Edité par François Norguet. VI, 306 pages. 1986.

Vol. 1189: J. Lukeš, J. Malý, L. Zajíček, Fine Topology Methods in Real Analysis and Potential Theory. X, 472 pages. 1986.

Vol. 1190: Optimization and Related Fields. Proceedings, 1984. Edited by R. Conti, E. De Giorgi and F. Giannessi. VIII, 419 pages. 1986.

Vol. 1191: A.R. Its, V.Yu. Novokshenov, The Isomonodromic Deformation Method in the Theory of Painlevé Equations. IV, 313 pages. 1986.

Vol. 1192: Equadiff 6. Proceedings, 1985. Edited by J. Vosmansky and M. Zlámal. XXIII, 404 pages. 1986.

Vol. 1193: Geometrical and Statistical Aspects of Probability in Banach Spaces. Proceedings, 1985. Edited by X. Femique, B. Heinkel, M.B. Marcus and P.A. Meyer. IV, 128 pages. 1986.

Vol. 1194: Complex Analysis and Algebraic Geometry. Proceedings, 1985. Edited by H. Grauert. VI, 235 pages. 1986.

Vol.1195: J.M. Barbosa, A.G. Colares, Minimal Surfaces in $\mathbb{R}^3$. X, 124 pages. 1986.

Vol. 1196: E. Casas-Alvero, S. Xambó-Descamps, The Enumerative Theory of Conics after Halphen. IX, 130 pages. 1986.

Vol. 1197: Ring Theory. Proceedings, 1985. Edited by F.M.J. van Oystaeyen. V, 231 pages. 1986.

Vol. 1198: Séminaire d'Analyse, P. Lelong – P. Dolbeault – H. Skoda. Seminar 1983/84. X, 260 pages. 1986.

Vol. 1199: Analytic Theory of Continued Fractions II. Proceedings, 1985. Edited by W.J. Thron. VI, 299 pages. 1986.

Vol. 1200: V.D. Milman, G. Schechtman, Asymptotic Theory of Finite Dimensional Normed Spaces. With an Appendix by M. Gromov. VIII, 156 pages. 1986.

Vol. 1201: Curvature and Topology of Riemannian Manifolds. Proceedings, 1985. Edited by K. Shiohama, T. Sakai and T. Sunada. VII, 336 pages. 1986.

Vol. 1202: A. Dür, Möbius Functions, Incidence Algebras and Power Series Representations. XI, 134 pages. 1986.

Vol. 1203: Stochastic Processes and Their Applications. Proceedings, 1985. Edited by K. Itô and T. Hida. VI, 222 pages. 1986.

Vol. 1204: Séminaire de Probabilités XX, 1984/85. Proceedings. Edité par J. Azéma et M. Yor. V, 639 pages. 1986.

Vol. 1205: B.Z. Moroz, Analytic Arithmetic in Algebraic Number Fields. VII, 177 pages. 1986.

Vol. 1206: Probability and Analysis, Varenna (Como) 1985. Seminar. Edited by G. Letta and M. Pratelli. VIII, 280 pages. 1986.

Vol. 1207: P.H. Bérard, Spectral Geometry: Direct and Inverse Problems. With an Appendix by G. Besson. XIII, 272 pages. 1986.

Vol. 1208: S. Kaijser, J.W. Pelletier, Interpolation Functors and Duality. IV, 167 pages. 1986.

Vol. 1209: Differential Geometry, Peñíscola 1985. Proceedings. Edited by A.M. Naveira, A. Ferrández and F. Mascaró. VIII, 306 pages. 1986.

Vol. 1210: Probability Measures on Groups VIII. Proceedings, 1985. Edited by H. Heyer. X, 386 pages. 1986.

Vol. 1211: M.B. Sevryuk, Reversible Systems. V, 319 pages. 1986.

Vol. 1212: Stochastic Spatial Processes. Proceedings, 1984. Edited by P. Tautu. VIII, 311 pages. 1986.

Vol. 1213: L.G. Lewis, Jr., J.P. May, M. Steinberger, Equivariant Stable Homotopy Theory. IX, 538 pages. 1986.

Vol. 1214: Global Analysis – Studies and Applications II. Edited by Yu.G. Borisovich and Yu.E. Gliklikh. V, 275 pages. 1986.

Vol. 1215: Lectures in Probability and Statistics. Edited by G. del Pino and R. Rebolledo. V, 491 pages. 1986.

Vol. 1216: J. Kogan, Bifurcation of Extremals in Optimal Control. VIII, 106 pages. 1986.

Vol. 1217: Transformation Groups. Proceedings, 1985. Edited by S. Jackowski and K. Pawalowski. X, 396 pages. 1986.

Vol. 1218: Schrödinger Operators, Aarhus 1985. Seminar. Edited by E. Balslev. V, 222 pages. 1986.

Vol. 1219: R. Weissauer, Stabile Modulformen und Eisensteinreihen. III, 147 Seiten. 1986.

Vol. 1220: Séminaire d'Algèbre Paul Dubreil et Marie-Paule Malliavin. Proceedings, 1985. Edité par M.-P. Malliavin. IV, 200 pages. 1986.

Vol. 1221: Probability and Banach Spaces. Proceedings, 1985. Edited by J. Bastero and M. San Miguel. XI, 222 pages. 1986.

Vol. 1222: A. Katok, J.-M. Strelcyn, with the collaboration of F. Ledrappier and F. Przytycki, Invariant Manifolds, Entropy and Billiards; Smooth Maps with Singularities. VIII, 283 pages. 1986.

Vol. 1223: Differential Equations in Banach Spaces. Proceedings, 1985. Edited by A. Favini and E. Obrecht. VIII, 299 pages. 1986.

Vol. 1224: Nonlinear Diffusion Problems, Montecatini Terme 1985. Seminar. Edited by A. Fasano and M. Primicerio. VIII, 188 pages. 1986.

Vol. 1225: Inverse Problems, Montecatini Terme 1986. Seminar. Edited by G. Talenti. VIII, 204 pages. 1986.

Vol. 1226: A. Buium, Differential Function Fields and Moduli of Algebraic Varieties. IX, 146 pages. 1986.

Vol. 1227: H. Helson, The Spectral Theorem. VI, 104 pages. 1986.

Vol. 1228: Multigrid Methods II. Proceedings, 1985. Edited by W. Hackbusch and U. Trottenberg. VI, 336 pages. 1986.

Vol. 1229: O. Bratteli, Derivations, Dissipations and Group Actions on C*-algebras. IV, 277 pages. 1986.

Vol. 1230: Numerical Analysis. Proceedings, 1984. Edited by J.-P. Hennart. X, 234 pages. 1986.

Vol. 1231: E.-U. Gekeler, Drinfeld Modular Curves. XIV, 107 pages. 1986.

Vol. 1232: P.C. Schuur, Asymptotic Analysis of Soliton Problems. VIII, 180 pages. 1986.

Vol. 1233: Stability Problems for Stochastic Models. Proceedings, 1985. Edited by V.V. Kalashnikov, B. Penkov and V.M. Zolotarev. VI, 223 pages. 1986.

Vol. 1234: Combinatoire énumérative. Proceedings, 1985. Edité par G. Labelle et P. Leroux. XIV, 387 pages. 1986.

Vol. 1235: Séminaire de Théorie du Potentiel, Paris, No. 8. Directeurs: M. Brelot, G. Choquet et J. Deny. Rédacteurs: F. Hirsch et G. Mokobodzki. III, 209 pages. 1987.

Vol. 1236: Stochastic Partial Differential Equations and Applications. Proceedings, 1985. Edited by G. Da Prato and L. Tubaro. V, 257 pages. 1987.

Vol. 1237: Rational Approximation and its Applications in Mathematics and Physics. Proceedings, 1985. Edited by J. Gilewicz, M. Pindor and W. Siemaszko. XII, 350 pages. 1987.

Vol. 1238: M. Holz, K.-P. Podewski and K. Steffens, Injective Choice Functions. VI, 183 pages. 1987.

Vol. 1239: P. Vojta, Diophantine Approximations and Value Distribution Theory. X, 132 pages. 1987.

Vol. 1240: Number Theory, New York 1984–85. Seminar. Edited by D.V. Chudnovsky, G.V. Chudnovsky, H. Cohn and M.B. Nathanson. V, 324 pages. 1987.

Vol. 1241: L. Gårding, Singularities in Linear Wave Propagation. III, 125 pages. 1987.

Vol. 1242: Functional Analysis II, with Contributions by J. Hoffmann-Jørgensen et al. Edited by S. Kurepa, H. Kraljević and D. Butković. VII, 432 pages. 1987.

Vol. 1243: Non Commutative Harmonic Analysis and Lie Groups. Proceedings, 1985. Edited by J. Carmona, P. Delorme and M. Vergne. V, 309 pages. 1987.

Vol. 1244: W. Müller, Manifolds with Cusps of Rank One. XI, 158 pages. 1987.

Vol. 1245: S. Rallis, L-Functions and the Oscillator Representation. XVI, 239 pages. 1987.

Vol. 1246: Hodge Theory. Proceedings, 1985. Edited by E. Cattani, F. Guillén, A. Kaplan and F. Puerta. VII, 175 pages. 1987.

Vol. 1247: Séminaire de Probabilités XXI. Proceedings. Edité par J. Azéma, P.A. Meyer et M. Yor. IV, 579 pages. 1987.

Vol. 1248: Nonlinear Semigroups, Partial Differential Equations and Attractors. Proceedings, 1985. Edited by T.L. Gill and W.W. Zachary. IX, 185 pages. 1987.

Vol. 1249: I. van den Berg, Nonstandard Asymptotic Analysis. IX, 187 pages. 1987.

Vol. 1250: Stochastic Processes – Mathematics and Physics II. Proceedings 1985. Edited by S. Albeverio, Ph. Blanchard and L. Streit. VI, 359 pages. 1987.

Vol. 1251: Differential Geometric Methods in Mathematical Physics. Proceedings, 1985. Edited by P.L. García and A. Pérez-Rendón. VII, 300 pages. 1987.

Vol. 1252: T. Kaise, Représentations de Weil et GL_2 Algèbres de division et GL_n. VII, 203 pages. 1987.

Vol. 1253: J. Fischer, An Approach to the Selberg Trace Formula via the Selberg Zeta-Function. III, 184 pages. 1987.

Vol. 1254: S. Gelbart, I. Piatetski-Shapiro, S. Rallis. Explicit Constructions of Automorphic L-Functions. VI, 152 pages. 1987.

Vol. 1255: Differential Geometry and Differential Equations. Proceedings, 1985. Edited by C. Gu, M. Berger and R.L. Bryant. XII, 243 pages. 1987.

Vol. 1256: Pseudo-Differential Operators. Proceedings, 1986. Edited by H.O. Cordes, B. Gramsch and H. Widom. X, 479 pages. 1987.

Vol. 1257: X. Wang, On the C*-Algebras of Foliations in the Plane. V, 165 pages. 1987.

Vol. 1258: J. Weidmann, Spectral Theory of Ordinary Differential Operators. VI, 303 pages. 1987.